The HISTORY and DEVELOPMENT of SMALL ARMS AMMUNITION
Volume I – Martial Long Arms, Flintlock Through Rimfire

Group of Harquebusiers from "Exercise of Armes For Calivres, Muskettes, and Pikes," Jacob DeGheyn, 1607 after a 1560 book by Jost Amman.

". . . The Harquebusse hath bin inuented within these fewe yeares, and is verie good, so that it be vsed by those that haue skill, but at this present euery man will be a Harquebusier: I know not whether it be to take the more wages, or to be the lighter laden, or to fight the further off, wherein there must be an order taken to appoint fewer Harquebusiers, and those that are good, than many that are worth nothing: for this negligence is cause that in a skirmish wherein tenne thousand Harquebussados are shot, there dieth not so mutch as one man, for the Harquebusiers content themselves with making of a noyse, and so shoot at all aduenters. . ."

From "Instructions for the Warres," de Fourquevaux, 1589 as quoted by Hon. T.F. Fremantle in "The Book of the Rifle," 1901

The History and Development *of* SMALL ARMS AMMUNITION

George A. Hoyem

Volume One

Martial Long Arms: Flintlock through Rimfire

1981

Armory Publications: Tacoma

This book is dedicated to a distinguished company of 19th and 20th century authors and editors, working from 1855 to the present, whose contributions to the story of small arms ammunition inspired this work. They are:

Capt. J. Schon (Royal Saxon Infantry); Maj. Alfred Mordecai (United States Army); Lt. C.M. Wilcox (United States Army); Maj. Rudolf Schmidt (Swiss Army); A. Mattenheimer, Hermann Weygand, Anton Dolleczek, Moritz Thierbach, Col. Berkeley R. Lewis (United States Army), Fred A. Datig, Charles R. Suydam. The late Frank Wheeler, former cartridge editor, Gun Report magazine; A. Donald Amesbury, retired secretary, International Cartridge Collectors Association; Emile Timmermans, retired secretary, European Cartridge Collectors Club.

Typography by De Fremery & Co.
547 Howard Street
San Francisco, California 94105

Printed in the United States of America by
Braun-Brumfield, Inc.
100 N. Staebler Road
Ann Arbor, Michigan 48106

Library of Congress Catalog Card Number: 80-67532

ISBN-0-9604982-8-1

First Printing

ACKNOWLEDGMENTS

That this book could be brought together at all is a tribute to the generosity of the collectors who have lent their support to me during the past three years.

Particularly I want to thank Gene Spicer, Paul Parsons, R.L. (Randy) Long, R. Peyton Hall, George Kass, Victor v.B. Engel, J.A. Harrold, E.L. (Gene) Scranton, and Steve Brenger for photographs of cartridges and rifles in their collections.

Howard Madaus, Curator of the Nunnemacher Arms Collection, Milwaukee Public Museum, and C.R. Goins, Curator, Division of Military History, Smithsonian Institution, provided photographs of certain American arms and experimental cartridges.

Hr. Arne Orloff, Chief Curator, Tøjhusmuseet (National Museum), Copenhagen, Denmark, together with Hr. Frederik Hasle, arranged for photographs of Scandinavian military arms.

Andrew Hoyem, San Francisco printer, gave valuable assistance in connection with design and publication of the book. Robert Hoyem, Kiel, Germany, opera state director, lent a generous hand in translating some German source material. Several of the people whose names appear here took time to read and criticize the manuscript. However, the contents as presented are my own responsibility.

My sincerest thanks go to each of the people named below who contributed significantly to Volume I.

Vidar Andresen (Norway)
John A. Belton (Canada)
John P. Blake
Jack A. Brickell
Tom and Stella Dunn
Henry U. Fuller
Ray Hanning
Ralph Hemstalk
Bob Hill
Dr. Albert G. Hoyem
Paul Klatt
Richard A. King (illustrator)
Richard D. King
Terry L. Koster
Maj. P.J.F. Mead (England)
George H. Murphy
Robert Stevens
James C. Tillinghast
Dean Thomas
Emile Timmermans (Belgium)
M.E. Whiteley (illustrator)
W.H. (Bill) Woodin
C.W. (Chuck) Young

Cartridge and Box Photo Credits

Eugene D. Spicer:	FL 1 through 8, 10; FLR 1; PSB 4, 5; PR 5, 6, 7, 11 through 15, 21; PBL 5, 15, 20, 21, 27, 28, 32, 34.
R.L. Long:	PSB 1; PR 10(a), 27 through 31, 33; PBL 7(b), 8, 16, 17, 31, 43.
George Kass:	RFR 5, 12, 15, 16, 68, all box labels, p. 137 and appendix.
Victor v.B. Engel	RFR 6, 7(d and e), 8(b), 18 through 28, 30(a), 35(d), 36, 38 through 54, 57, 58, 59, 84; RFMG 2.
E.L. Scranton:	PBL 13

Other cartridge and box label photographs were taken by the compiler from his own and the collections of other contributors.

CONTENTS

Foreword

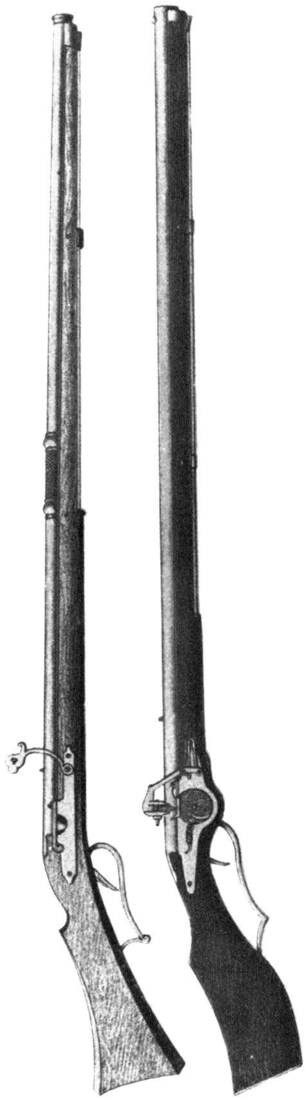

This first volume traces the development of ammunition for the military musket, rifle, and primitive machinegun, worldwide, from the 17th century when round lead ball and black powder were first wrapped together in a paper cylinder, through the first self-contained, metallic cartridge found practical for military use—the rimfire.

Volume II, already well along, will carry the study forward through the black powder era to cover the early centerfire rifle and machinegun cartridges as well as some of the first experiments in caliber reduction.

As the reader will note, the various developments in cartridges, and the arms which fired them, overlap each other chronologically. Prussian soldiers were armed with a breech-loading rifle and self-contained ammunition long before most countries abandoned the muzzle-loader. Percussion breech-loading rifles and self-consuming ammunition for them appeared in the early 1840's, but muzzle-loaders remained the primary infantry arms in most nations until the 1860's.

Accordingly, this book is organized by ammunition and weapons types rather than on a strict chronological basis.

As I progressed through the first chapters, more and more fascinating specimens of paper and combustible ammunition turned up for study. However, the line had to be drawn somewhere, so specimens have been left out if they tended to duplicate others in type or appearance.

Pistol and revolver cartridges have not been included except to illustrate important advancements. We already have good books on that subject (See Bibliography). I have shown a few of the early sporting rimfire cartridges which grew out of military loadings but commercial loadings of the rimfires developed for military use have been excluded where possible.

This book was inspired by my constant frustration at trying to match up a cartridge with its intended firearm, and vice versa. Hopefully, my efforts here will be of service to others who have experienced similar difficulties.

Historical Background

Gunpowder, or black powder as we know it today, was first produced during the 13th century in Europe following the discovery of the compound, saltpetre. Roger Bacon described a mixture of 7 parts saltpetre, 5 of charcoal, and 5 of sulphur which would explode when brought in contact with fire. Incendiary mixtures were known long before this in Europe and Asia, but there seems to be no certain evidence that they were used for propelling projectiles in those earlier times.

Ancient writings establish that cannon were in use in Europe by the early 1300's, and hand-held firearms appeared very soon thereafter. The hand cannon was a simple tube of cast bronze, brass, or wrought iron strapped to a length of wood or attached to a shaft. A touchhole, or vent, at the top of the breech led into the powder chamber.

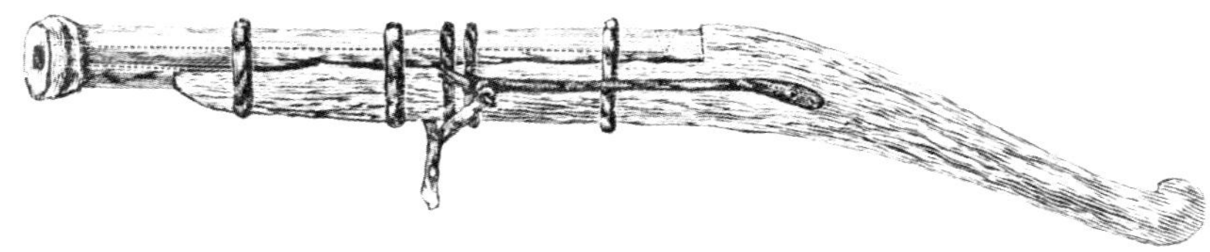

Primitive handcannon, or "culverin."

To fire this weapon, the soldier poured a charge of ground up, or "meal," gunpowder down the tube and followed it with a length of wadding or soft wood. Then he dropped a ball of stone, iron, brass, or lead down the tube on top of the wad and powder charge. Finally, he applied a small quantity of the gunpowder to the vent and let it pile up on the barrel. The firer pointed the barrel in the direction of his enemy and applied a lighted coal or red hot piece of metal to the priming powder. The resulting explosion was transmitted down the vent into the chamber to fire the main charge. The projectile was hurled out the open end of the tube by this explosion, hopefully to strike the intended target.

This procedure was inconvenient and lacking in precision. The soldier could not touch the coal or metal to the priming charge without taking his eyes off the enemy. Therefore, an improvement was devised. The vent was moved to the right side of the breech and a small pan added beneath it to hold the priming powder. This allowed the soldier to maintain some sort of aim while he applied fire to the pan with his right hand.

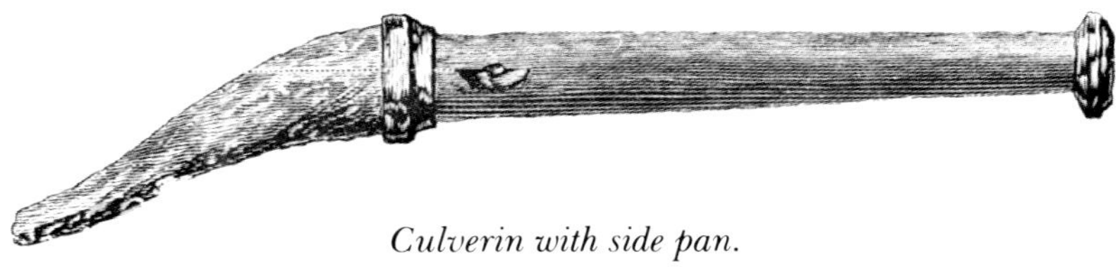

Culverin with side pan.

Eventually the awkward coal and hot iron igniting devices were replaced by the "match,"—a length of twisted cord wound with thread and lighted at one end.

The Matchlock

The hand cannon, or "culverin," went through many stages of gradual improvement, but the most important was the appearance of the matchlock sometime between 1375 and 1440. This was a mechanical device which would hold the lighted match and apply it to the pan while the soldier maintained his aim. In the most primitive types, a curved piece of metal was fastened to the side of the barrel behind the vent and flash pan. The "serpentine," as it was called, could rotate about the fastening pin in its center. The upper end, or head, was split lengthwise to form a clamp in which the match was held. When the lower arm of the serpentine was drawn to the rear in trigger fashion, the head came down and put the glowing match in contact with the priming powder in the pan. Ultimately, fine-ground priming powder came to be known as serpentine.

Primitive Matchlock Arm

These primitive firearms were hardly more effective than the longbow and crossbow, both of which continued to be used for military purposes long after firearms came into general military use. In the 1600's it was considered reasonable performance if the musketeer got off six or seven shots during the course of an eight-hour battle. The glowing match was a hazard in the presence of loose gunpowder. Carrying the lighted match on the march was detailed to one soldier in a small unit, and he lit those of his comrades before the battle. During rainy weather the match had to be carefully shielded.

Origins of the Cartridge

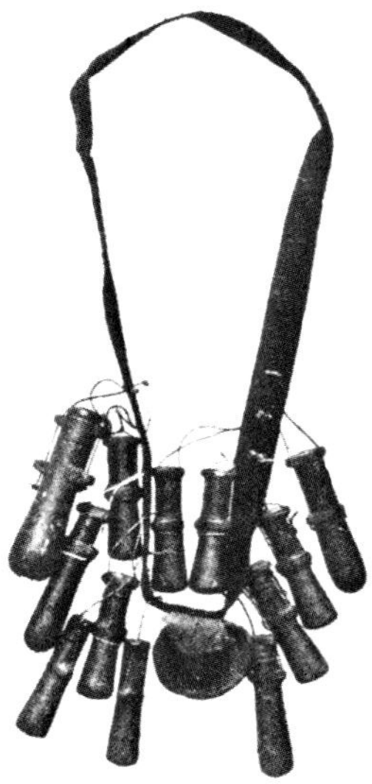

One of the first devices used to carry pre-measured powder charges for the musket was a hollow wooden capsule fitted with a lid. This idea emerged during the 1500's. A dozen of these, covered with leather to prevent them from rattling against each other, were strung on leather thongs attached to a bandolier which was carried over the shoulder. The lids, made with eyeholes on each side, were also strung on the thongs to prevent their loss. One of the capsules was normally used to carry the serpentine powder. The musketeer carried his lead ball projectiles in a separate leather bag, as well as a length of ignition match and flasks containing supplies of priming and musket powder.

The Wheel-lock, Snaphaunce, and Miquelet

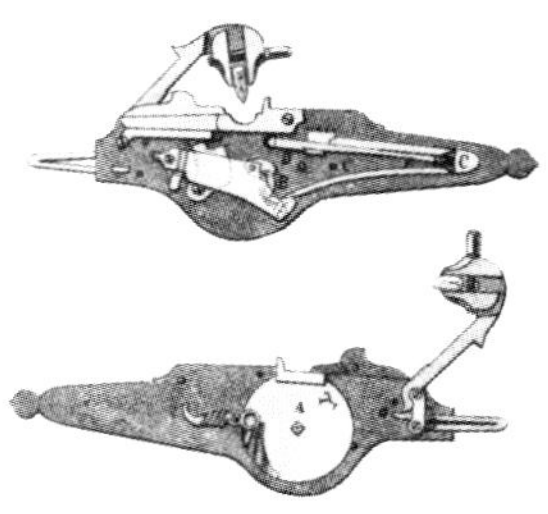

The Wheel-lock

The Snaphaunce Lock

The matchlock musket remained the primary firearm of the foot soldier during the 16th century. The wheel-lock, invented in Germany around 1500, was complicated and expensive to make and saw only limited military use. However, its function of producing a shower of sparks by striking iron pyrites against a revolving steel wheel to ignite a priming charge led to the appearance of the snaphaunce lock around 1525. Although it is usually attributed to the Spanish, the name emanates from the Dutch, evidently because the hammer holding stone in its jaws descending against the anvil in the pan reminded them of a pecking hen. The snaphaunce, and an improvement called the miquelet, were adopted by the principle military powers of the day and gradually began to replace the matchlock.

The Flintlock

The French flintlock developed between 1610 and 1630 incorporated the pan cover and frizzen, which were separate parts in the snaphaunce, into one piece. Flint replaced the iron pyrites and was clamped in the jaws of the hammer within a lining, usually leather. The frizzen was slanted sharply upward at just the right angle so that the hammer descending with the flint would knock it forward opening the pan cover and creating a shower of sparks from the flint scraping steel. The sparks fell into the pan, igniting the priming powder which set off the charge in the barrel.

These brief statements do not do justice to the fascinating subject of early firearm locks. However, for our purpose here, it will suffice to say that striking flint against steel to produce sparks remained the principle method of igniting priming powder to discharge a military firearm from the late 1600's until about 1840.

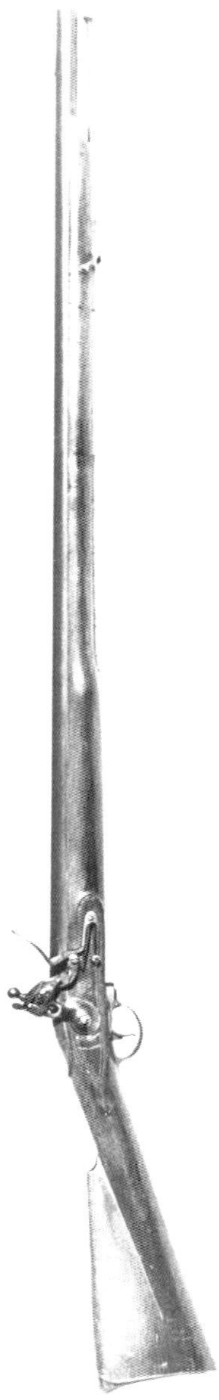

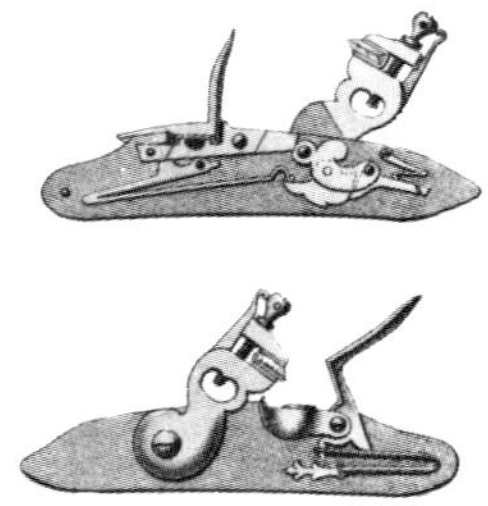

(Drawings from W. W. Greener's "The Gun and Its Development," 7th ed., 1899, and Rodolphe Schmidt's "Les Armes A Feu Portatives," 1877.)

The English Brown Bess flintlock musket.

1 THE CARTRIDGE FOR THE FLINTLOCK MUSKET

The soldier armed with smoothbore flintlock musket was trained to make his own ammunition. He used a cylindrical stick with a hollowed end in which he placed the ball. He then laid the stick, which was the diameter of the cartridge, upon a trapezium-shaped piece of unsized paper which he rolled around the stick and ball. He then folded the ball end of the paper closed and removed the stick. The powder charge was poured into the other end of the paper tube and that end was closed with a twist of the remaining paper which was folded along the body of the cartridge. The projectile end of the completed cartridge was either pasted or tied with string.

Contents of a paper cartridge for muzzle-loader—black powder and lead ball.

Loading the Musket

The musket was loaded from the standing position. The soldier bit off the tail of the cartridge and poured a small quantity of the exposed powder into the pan which he then closed. The remaining charge was poured down the barrel and the ball dropped in and rammed down on top of it with the wooden ramrod. The ball could be loaded naked or in its paper covering, depending on how fouled the bore might be with black powder residue. If the ball was loaded without covering, and was loose in the bore, the cartridge paper could be rammed down on top of it to keep it from working away from the powder or rolling back out the muzzle.

When the soldier fired his musket, the rapidly expanding gases resulting from the explosion of the powder escaped to some extent by the ball which was smaller than the interior diameter of the barrel. The projectile, propelled by the force of the explosion, tended to bound from one side of the bore to the other until it exited at the muzzle. Needless to say, this resulted in poor accuracy. However, accuracy was of little concern to military leaders of the time. The soldier was expected only to aim in the direction of a body of enemy troops at close range. The clouds of black powder smoke left by discharging muskets on both sides further reduced the chance of carefully aimed fire at individuals. The soldier was even provided with buckshot loads, often called grapeshot, containing several small projectiles which would obtain more hits at close range.

FL 1 – 18.3 mm (.72in) Austrian Infantry Musket, ball load, circa 1700

FL 2 – 18.3mm (.72in) Austrian Infantry Musket, three-ball or "grapeshot" load, circa 1700

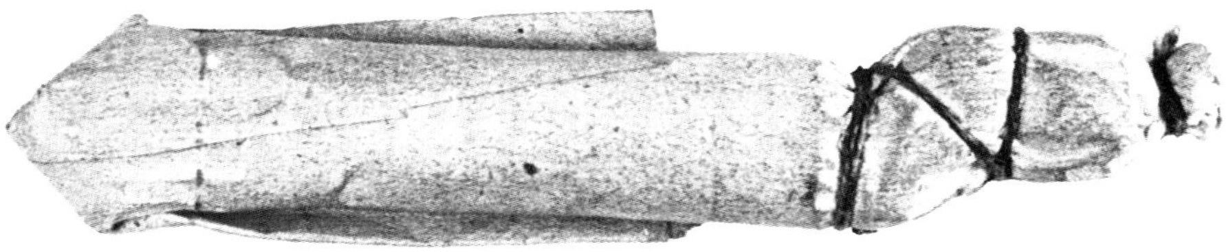

Identification

Unsized paper tube, tan or white. The paper case is attached to the ball load by a string tie around the sprue (lead adhering to round ball after moulding left on for that purpose). In the grapeshot load, the paper case is tied over the three balls equivalent in weight to the larger projectile with half hitches over the center and next to the powder. The tail of the case in both specimens is pressed flat, folded at a right angle, and laid along the body of the case.

Comments

These 280-year-old Austrian musket cartridges contain 200 and 185 grains of powder respectively. Such massive charges were necessary in the early flintlock period because of the relatively inferior powder which was only one-half or two-thirds as strong as that developed during the early 19th century. Moreover, 10 to 15 grains had to be used to prime the ignition pan before the main part of the charge was poured down the barrel.

FL 3 — 18.3mm (.72in) Austrian Infantry
Musket, ca. 1770

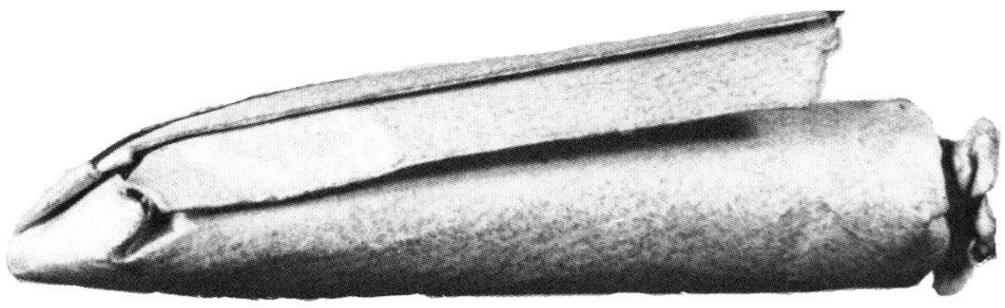

Description

Tan paper tube tied over a round ball with powder filled in contact with ball. Tail twisted and folded as before. Case length and powder charge have diminished.

FL 4 — 17.6mm (.69in) Austrian Infantry
Musket, ca. 1821

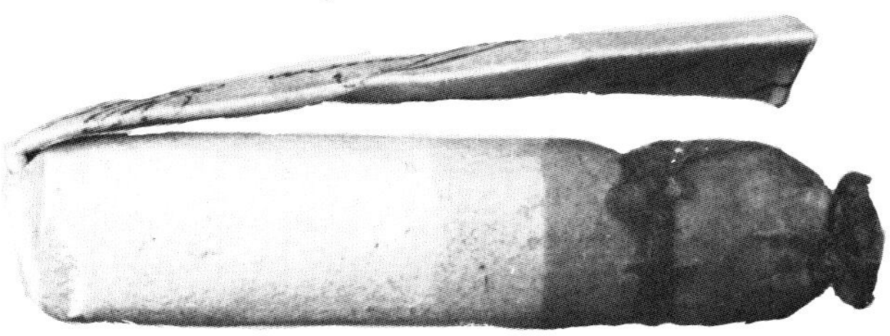

White paper tube stained light brown over ball tied over and below the ball. Ball lubricated with tallow, beeswax, or a combination thereof. Usual long tail folded at right angle and along the case body.

Comments

In addition to reducing the powder charge, the Austrians eventually reduced the bore diameter of the infantry musket. It remained the 17.6mm (.69in) used by most other nations until the end of the flintlock period. A refinement in cartridge manufacture evident in FL 4 is the tie below the ball which kept powder grains from working up around the ball. When this occurred in the more primitive cartridge, the cartridge diameter at that point was increased, making it difficult to get the projectile in its paper covering down the bore. The lubrication of the ball further eased loading and cleaned the bore of powder fouling. This feature would suggest that this cartridge was arsenal made, not prepared by a soldier in the field.

Austria, alone, introduced no less than 45 different models of flintlock muskets and carbines for its various military forces between 1700 and 1828. Bore sizes ranged from 18.3mm to 13.9mm. Consequently, many different cartridges were made for them over that span of time.

Reference: "Monographie der k.u.k. ŏsterr.-ung. blanken und Handfeuer-Waffen, Anton Dolleczek, 1896 (reproduced in 1970 by Akademische Druck-u. Verlagsanstalt, Graz, Austria.)

FL 5 – 15.6mm (.61in) Austrian Cavalry Musket, ca. 1802-1814

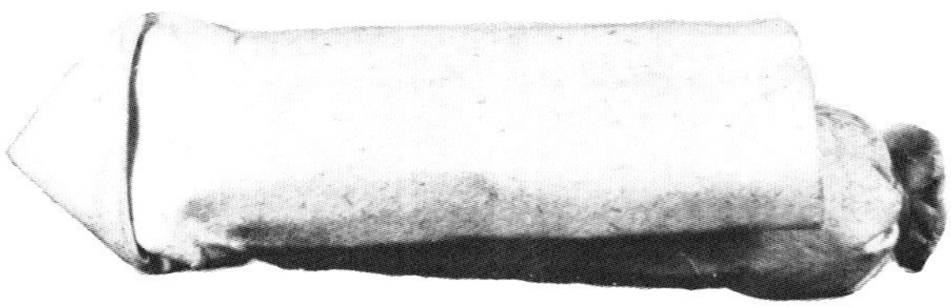

Description

Sandy-white paper case tied over the round ball. Otherwise, the same construction as FL 3. A reduced powder charge for the cavalry musket resulted in a shorter cartridge.

FL 6 – 17.53mm (.69in) French Charleville Infantry Musket, ca. 1776

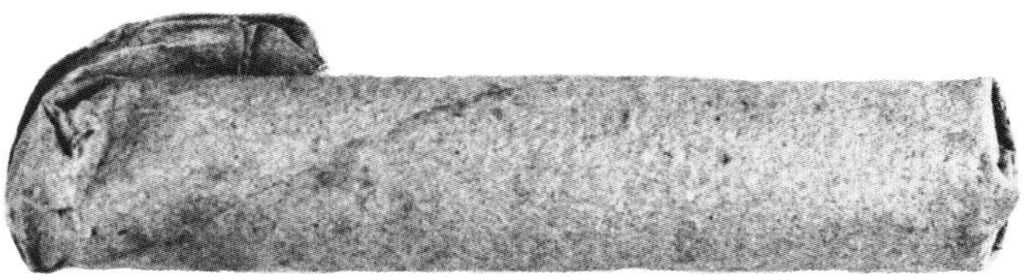

Tan unsized paper case pasted closed over the round ball. Closure of the powder column effected by twisting the tail and folding it along the case.

Comments

American revolutionary forces purchased over 100,000 of the French Charleville muskets and manufactured similar arms during the war with England. Consequently, ammunition for the .69 caliber smoothbores is the most common flintlock type to survive today.

FL 7 – .69in (17.53mm) U.S. Musket, ca. 1812

Description

Tan paper tube pasted over a round ball, powder filled on top of projectile and tail twisted closed.

FL 8 – .69in (17.53mm) U.S. Musket, ca. 1816

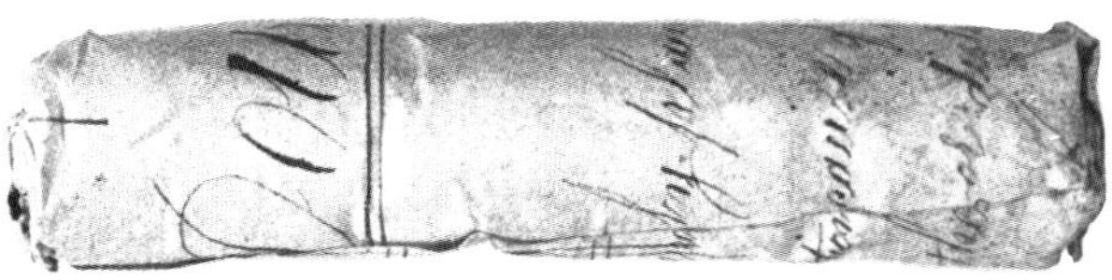

Yellowish white paper tube made of used ledger paper pasted closed over a round ball with powder charge filled in on top. Tail twisted closed. Around the case are the words: "Double. . . by placing the three. . . cause of gain, loss, or action. . . or distance of place, the . . ."

Comments

Both of the specimens above appear to be handmade by soldiers in the field. Note the straight seam on both cases. This case making procedure was inferior to rolling the paper at an angle to form a diagonal seam which was then tightened by folding or twisting the tail of the case.

Varying size balls will be found in such cartridges intended for the same musket because powder fouling after repeated shots sometimes required selecting a smaller projectile than normal to get it down the bore. Depending on the condition of the bore, the soldier could either load the single ball with paper still around it or shake it out of the paper and ram it home naked. The remaining paper could then be used as wadding over the ball to keep it against the charge.

FL 9 – .69in (17.53mm) U.S. Musket,
Buck and Ball, 1822-1840

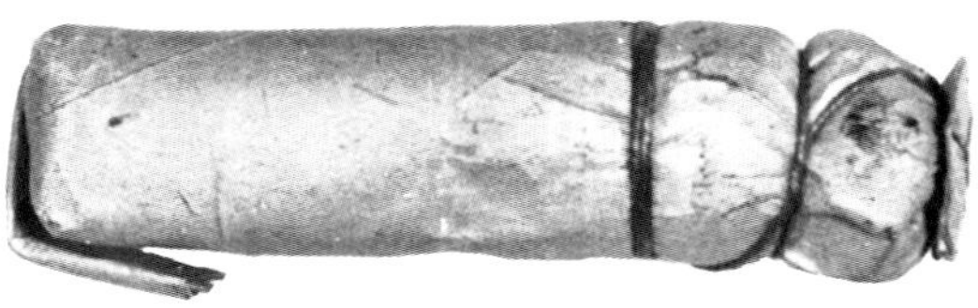

Description

Tan paper tube tied over three .31in shot loaded atop a regular size ball. String leads from the closure to form half hitches between the projectiles and below the large ball. Tail pressed flat, folded from the sides to the center, and laid along the case.

FL 10 – .62in (15.75mm) British Carbine,
ca. 1835-1845

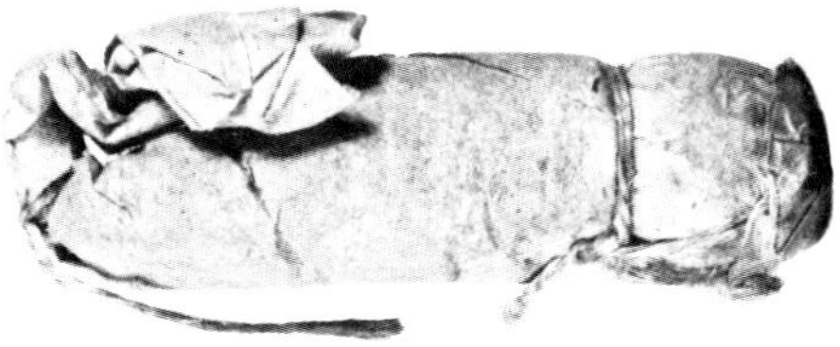

Light brown paper tube tied closed at the ball end with a half hitch tie below the ball. Powder column closed with string and the tail twisted along the case.

Comments

During the 1830's, the United States Army issued regulations which standardized loads for its infantry arms and specified cartridge making procedures. Ammunition made toward the end of the flintlock era reflects a reduced powder charge because powder quality had improved. Therefore, the powder train in the late loads is shorter than in the earlier loads. These cartridges are difficult to distinguish from ammunition made for the percussion smoothbores commencing in 1842. In fact, they could be loaded in either type arm.

The British carbine load shown could as easily have been used in a percussion smoothbore as a flintlock smoothbore. Its date as reflected by the White & Munhall collection records dictated that it be included in this section. British forces did not issue a percussion musket until 1841-1842.

In cartridges made with a string wrap between powder and ball, the projectile was normally loaded with its paper covering. The buck and ball load was always loaded with paper covering since the string ties were intended to keep the shot and ball charge together until out the muzzle.

2 THE ORIGINS OF RIFLING

The practice of cutting grooves screw-wise upon the interior surface of a firearm barrel was first applied by European gunmakers sometime in the 15th century. Rifles dated in the early 1500's are found in the museums of Vienna, Austria and Zurich, Switzerland. Crossbows had already been made to impart a stabilizing spin to the quarrel, or bolt. A spirally grooved musket bore which would grip a lead ball and give it a spinning motion advanced the same principle. The superior accuracy of the ball fired from a rifled bore led the Swiss to formulate regulations in 1563 to settle disputes between riflemen and shooters of the smoothbore which arose during match competition.

Because it required force to drive a lead ball down a rifled bore so that the projectile would engage and be spun by the rifling, loading such weapons was a slow, tedious process. The exertion required was so great that the shooter could not do proper justice to the precision of his rifle after he did get it loaded. Accordingly, military leaders found rifles impractical where battles were won by masses of troops delivering volumes of musket fire at close range rather than by accurate shooting at a distance.

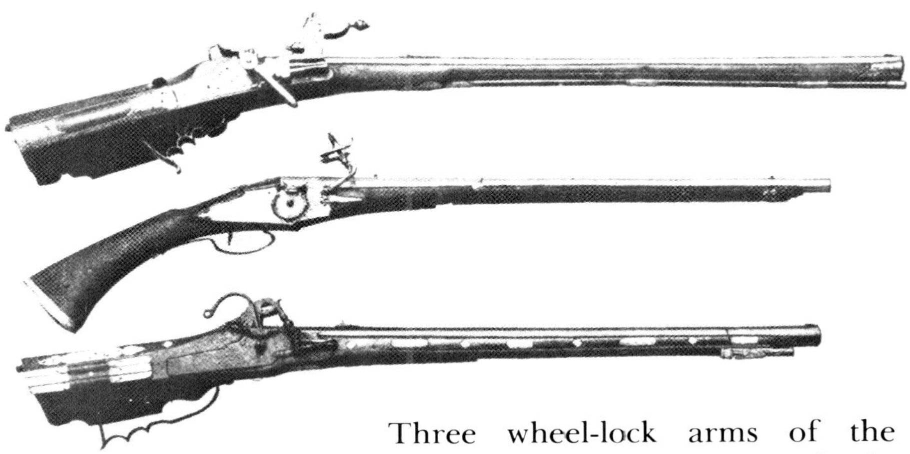

Three wheel-lock arms of the 1600's, the upper two made in Germany, the lower in Italy. The upper and lower specimens have rifled barrels. (From the Nunnemacher collection, Milwaukee Public Museum)

Early Military Experiments

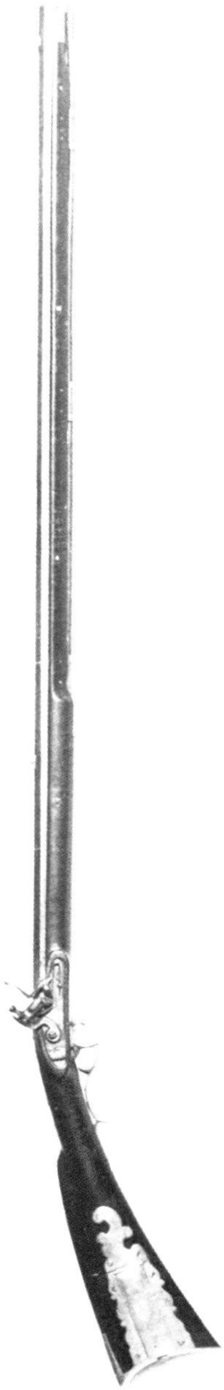

Cal. .50 Flintlock Pennsylvania Long Rifle, ca. 1800

Despite its drawbacks as a general issue weapon, Elector Maximilian of Bavaria and Frederic William of Prussia introduced the rifle to their ground armies for limited use in special sharpshooter units. About the same time, Louis XIV armed some of the French cavalry with rifled carbines. The designation "carbine" in those days denoted a shoulder firearm of smaller caliber than the infantry musket, not necessarily with shorter barrel. The English experimented with them, issuing eight rifled carbines to every company of the English Life Guards in 1680. In the late 1700's, both the Austrians and the French investigated the practicality of flintlock rifles. However, the Austrians withdrew theirs from service in 1801, and the French rifled Carbine Versailles adopted in 1793 was ordered withdrawn in 1805 by Napoleon. This weapon required a mallet and and stout ramrod to get the bullet down the barrel, and the Emperor described it as the worst weapon that could be got into the hands of a soldier.

British Army Experiences in late 1700's

English writers have acknowledged that events of the war with the American colonies in the late 18th century finally led the British army to adopt a rifled infantry arm for special units. The American insurrectionists, armed in part with the long-barreled, smallbore Pennsylvania rifle developed by German immigrants, raised havoc with British troops at long range. During the war, the British enlisted several regiments of German mercenaries armed with rifles to cope with the Americans, and also had several hundred "Kentucky" type rifled .54 caliber flintlocks made up at home for issue to these troops.

The First Military Breech-loading Rifle

One British officer, Lt. Col. Patrick Ferguson, predicted the ultimate supremacy of the rifle in military affairs and developed a screw-breech rifle modeled after similar sporting rifles. He armed his unit with this rifle during the American Revolution and demonstrated its merits before his superiors, getting off as many as six shots in a minute with notable accuracy under wet, windy weather conditions. However, the Ferguson rifle was not seriously considered by the British military for adoption. Ultimately, Ferguson was killed on October 6, 1780 at King's Mountain, North Carolina by American irregulars who surrounded, decimated, and captured the surviving members of his unit.

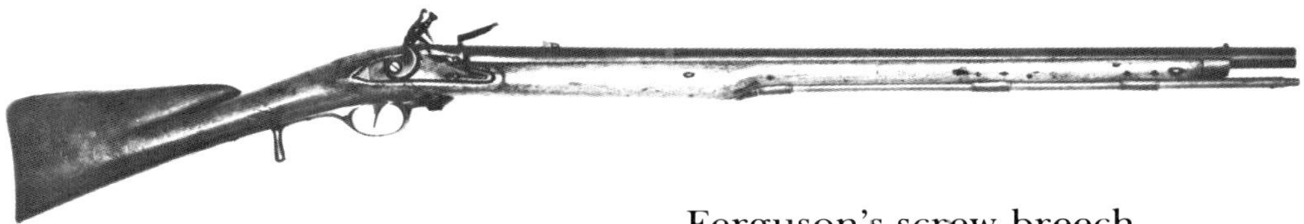

Ferguson's screw-breech rifle action which permitted loading powder and ball through an opening atop the breech.

The Baker Rifle

It was not until 1800 that Great Britain adopted a rifle made by one Ezekiel Baker of London. This .615in (15.62mm) arm still had the disadvantage at first of being difficult to load because a near bore-size ball was used. Eventually, the diameter of the ball was reduced and wrapped in a bit of greased linen. This eased loading, cleaned the bore, eliminated much of the windage, and assured that the rifling functioned to grip the ball wrapped in fabric. Of course, this very procedure had been developed long before on the European continent and brought to America by the Germans. The usefulness of the Rifle Brigade which was armed with the Baker was soon acknowledged by the British and it was still in limited service during the Crimean War, 1854-1856.

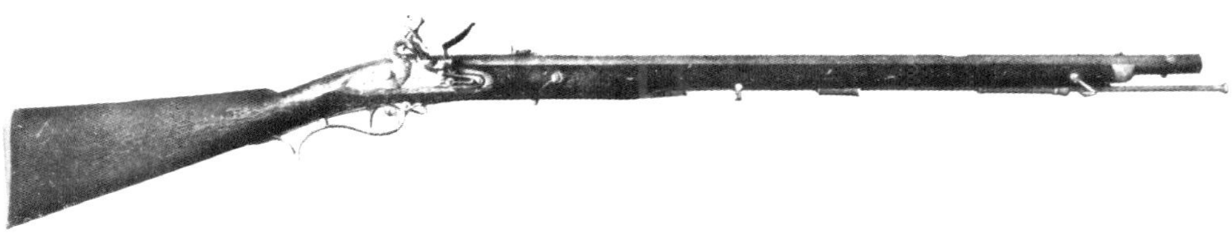

The Muzzle-loading Baker Flintlock rifle adopted by Great Britain in 1800.

Other European and American Developments

Among other rifle experiments, the Austrians tried a novel air rifle in 1779 which was claimed to be effective up to 120 yards and could be fired rapidly 20 times without recharging the reservoir. However, it required 2000 pump strokes to recharge a reservoir so spares had to be carried. This arm was withdrawn in 1801 and its manufacture subsequently forbidden. Its rapidity of fire probably upset military thinking.

In the newly-independent United States, a rifle battalion was authorized in 1792 and Pennsylvania gunsmiths made some flintlock rifles for it on contract. It was not until 1804 that a rifle regiment was authorized and manufacture of rifles commenced at Harpers Ferry Arsenal, Virginia. Several thousand of a newer model approved in 1814 were ordered from private makers. This was a half-stocked arm resembling the Pennsylvania rifles. Yet another muzzle-loading model appeared in 1819 with full stock.

That same year, the United States army commenced trials of a flintlock breech-loading rifle patented by Colonel John H. Hall in 1811. Its tip-up breech piece bore close resemblance to the Austrian Gorla-Crespi smoothbore flintlock breech-loader adopted by that country on a limited scale in 1771. The Hall rifle saw many years of service through the flintlock era and into the percussion era.

References: "The Book of the Rifle," Hon. T. F. Fremantle, 1901.
"Small Arms and Ammunition in the United States Service," Col. Berkeley R. Lewis, 1956.
"Pictorial History of the Rifle," G. W. P. Swenson, 1972.
"The Gorla/Crespi Carbine," F.H. Baer, Vienna, Austria, The Gun Report, Dec. 1979.

U.S. Model 1814 Harpers Ferry rifle dated 1815.(Smithsonian Institution photo—W.C. Dodge collection)

The American Hall flintlock breech-loader with hinged block which tipped up for loading. Introduced into the U.S. Army in 1819.

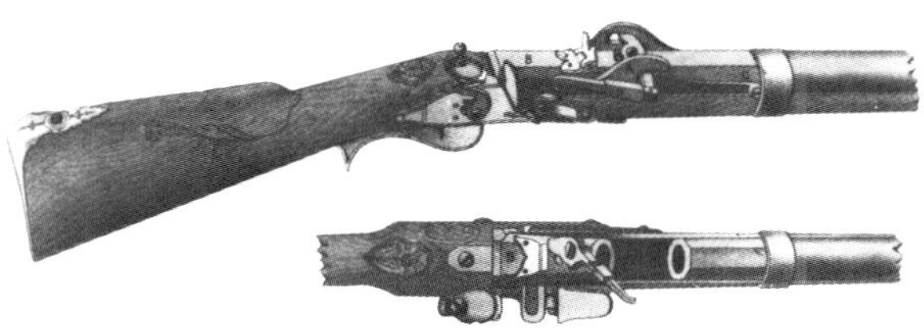

Austrian Gorla/Crespi smoothbore breech-loader with same type of flintlock action, adopted on a limited scale in that country in 1771. (Drawing from Rudolf Schmidt's "Die Handfeuerwaffen," 1875-1878)

The Patched Ball

The principle of using the patched ball was developed in Central Europe and came to the Americas in the 1700's via German immigrants where it was to become the standard method for loading the Pennsylvania long rifle. Actually, no cartridge was used with these rifles. The separate powder charge was loaded in the barrel and the greased linen or leather patch was laid centered on the muzzle. The ball was placed on it and patch and ball were rammed home, care being taken not to crush the powder. The patch wiped the bore clean, sealed the bore so that powder gases could not get by the ball, and allowed the projectile to "take" the rifling and spin as it traveled down the bore.

The same system was employed for the first official U.S. flintlock rifles. The ball was wrapped in a patch of bladder or other animal membrane. These were loaded in a paper cartridge similar to FL 9 shown previously but the soldier removed the paper from the ball before loading it.

The Hall flintlock employed a similar paper cartridge, but the lead ball was loaded bare. The chamber in the elevating breechblock was made larger than the bore so a ball larger than the bore proper could be used. Of course, when it struck the rifling it sealed the bore.

The English Ferguson rifle, as well as the Baker flintlock rifle, were both loaded with loose powder and ball. As noted, the Baker's ball was wrapped in greased linen. However, according to Major Myatt, the soldier equipped with the Baker carried regular musket cartridges for use during field emergencies when he had to be able to load and fire rapidly. With such ammunition, the Baker did not function as a rifle at all, because the smaller-than-bore ball used could not take the rifling.

References: "Small Arms and Ammunition in the United States Service," B. R. Lewis, 1956.
"The Illustrated Encyclopedia of Firearms," Major F. C. Myatt, MC, 1979.

FLR – 15.6mm (.61in) Austrian Cavalry Rifle, ca. 1790

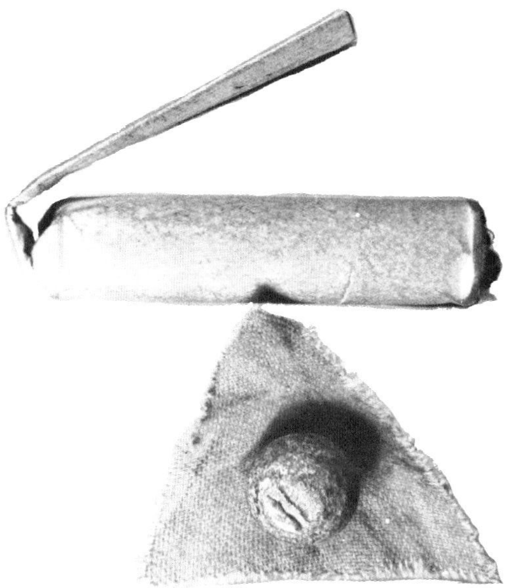

Description

Pale green tube for powder charge with lead ball and triangular brown linen patch carried separately.

Comments

The Austrians, according to Anton Dolleczek, were issuing rifles to sharpshooters as early as 1768. This specimen reflects the technique long established in Central Europe by 1790 of wrapping the ball in a greased linen patch. The patch facilitated loading, lubricated and cleaned the bore of fouling, and sealed the exploding powder gases behind the ball. The rifling acting on the patch gave the ball the spin which stabilized its flight and provided the accuracy necessary for shooting at longer than smoothbore musket range.

The specimen, once in the White & Munhall collection, was identified from those records. It is also illustrated in Dolleczek's "Monographie der k.u.k. osterr.-ung. blanken und Handfeuer-Waffer," 1896, reprinted in 1970 by Akademische Druck-u. Verlagsanstalt, Graz, Austria.

FLINTLOCK ARMS AND CARTRIDGE DATA

No.	Musket Bore	Model and Date	Ball Diameter	Ball Weight	Powder Charge	Cartridge Length	Sources
FL 1	18.3mm .72in	Austrian Inf., 1770	16.76mm .66in	30.6gm 468grs	13gm 200grs	117.35mm 4.62in	Specimen—Dolleczek
FL 2	18.3mm .72in	Austrian Inf., 1770	16.76mm .66in	30.6gm 468grs	12gm 185grs	117.35mm 4.62in	Specimen—Dolleczek
FL 3	18.3mm .72in	Austrian Inf., ca 1770	16.76mm .66in	30.6gm 468grs		92.1mm 3.62in	Specimen—W & M
FL 4	17.6mm .69in	Austrian Inf., ca 1821					
FL 5	15.6mm .61in	Austrian Cavalry, 1802				72.39mm 2.85in	Specimen—W & M
FL 6	17.53mm .69in	French, ca 1776	16mm .630in	23.8gm 368grs	12.3gm 189grs	87.63mm 3.45in	Specimen— Lewis
FL 7	.69in 17.53mm	American, 1812	.640in 16.26mm	389grs 25.2gm	160grs 10.4gm	78.74mm 3.1in	Specimen—Lewis
FL 8	.69in 17.53mm	American, ca 1816	.640in 16.26mm	389grs 25.2gm	160grs 10.4gm	3.15in 80mm	Specimen—Lewis
FL 9	.69in 17.53mm	American, ca 1822, 1840	.640in 16.26mm	389grs 25.2gm	160grs 10.4gm	2.55in 64.77mm	Specimen—Lewis
FL 10	.62in 15.75mm	British, ca 1835-1845	.58-60in 14.7-15mm			2.38in 60.5mm	Specimen
FLR 1	15.6mm .61in	Austrian, ca 1790	14.73mm .58in	21.8gm 336grs	5.5gm 84.9grs	63mm 2.48in	Specimen—Dolleczek

Note: The Austrian specimens were part of the White & Munhall collection and identifications were contained in their records. Verifications were made, where possible, by reference to Anton Dolleczek's "Monographie der k.u.k. osterr.-ung. blanken und Handfeuer-Waffen." Dimensions of projectiles, except for FLR 1, were taken over the paper. Actual diameter of the ball will be from .10 to .20 thousandths less, depending on thickness of paper.

3 THE PERCUSSION IGNITION SYSTEM

Through the first four hundred years of its use as a military weapon, the firearm was discharged by applying fire or sparks to black powder. The properties of fulminates (salts formed by the dissolution of metals in acids which explode when struck a violent blow,) had been known as early as the 1600's. It was not until 1807, however, that the Reverend Alexander J. Forsyth, a Scottish clergyman, patented a successful percussion lock which he fitted to a fowling piece.

Encouraged to develop a more solid system for military as well as sporting use, Rev. Forsyth devised a magazine often called the "scent bottle" attached to the lock which would tip and deposit a small amount of fulminate composition in a small pan adjacent to the vent, or flash hole. When the hammer dropped, it struck a small firing pin which in turn struck into the pan, exploding the fulminate and firing the weapon. Later devices employed the fulminate mixture formed into small pills or patches which were struck directly by the hammer nose. Other gunmakers devised plugs designed to take the priming mixture internally as well as priming tubes filled with fulminate which were inserted into the touchhole and crushed by the hammer to explode them.

The Percussion Cap

19th century percussion caps with embossed national emblems or makers' names.

One Joshua Shaw, an Englishman, was issued a patent in 1822 by the United States for what we still know as the percussion cap. He is popularly credited with the invention. However, two Frenchmen—Prelat and Deboubert—were issued patents in their country for the same idea in 1818 and 1820 respectively. Other experimenters also laid claim to the idea later on. Whoever actually invented the hollow copper cap closed at one end and containing fulminate mixture in its interior base cleared the way for rapid advancements in the construction of locks which would replace the old flintlock and eventually formed the essential basis for the breech-loader and self-contained ammunition.

The percussion cap was designed to be placed by hand onto the open end of a tube, or "nipple," screwed into an ignition orifice leading to the firearm's chamber.

Flintlocks Converted To Percussion

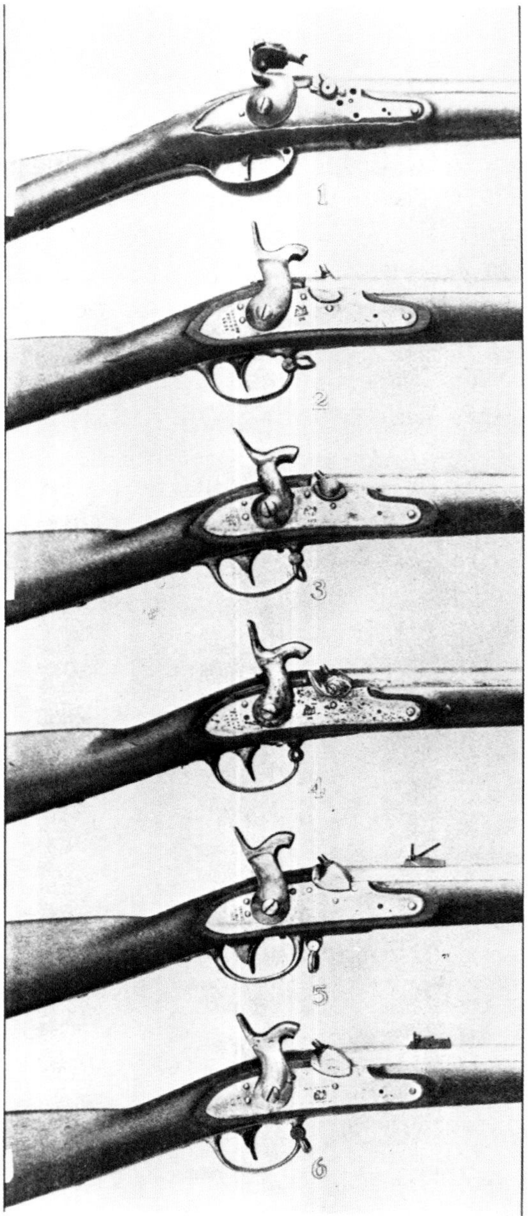

Description

1. A steel hammer nose replaced the flint in this English specimen. The original vent was enlarged and a cylindrical lug was screwed into it. Then the cone was screwed into a threaded hole in the side of the hollow lug.

2. A percussion hammer replaced the flintlock type. A cone was screwed into the barrel and the original vent plugged. The lock plate is stamped "US" and "Harpers Ferry 1831."

3. "US Springfield 1829" is stamped on the lock of this specimen. A percussion hammer replaced the original flint and the lug and cone are essentially the same as in No. 1 above.

4. "US Harpers Ferry 1839." The old breech was cut off and a new one screwed into the threaded rear of the barrel. The new breech has a nipple cone extending over the edge of the lock plate.

5. Lock plate is marked "US Springfield 1818." The old breech was replaced as in No. 4 above, new one incorporating a nipple stud on the side over the lock plate.

6. Hughes & Phillips of Newark, N.J. converted this flintlock, U.S. 1822 model musket in 1862 in the same manner as No. 5.

(From the original 1928 catalogue of the Nunnemacher Arms Collection, courtesy, Milwaukee Public Museum.)

Percussion Tubes, Tapes, and Discs

Description

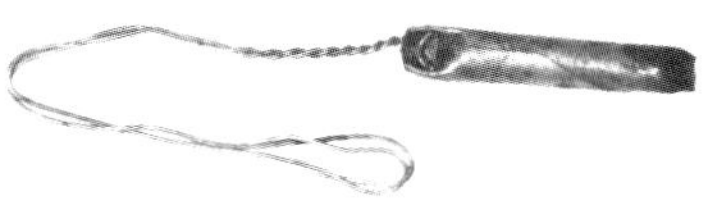

Console tube ignitor made of brass foil rolled into a tube, pinched and cemented closed at one end and pierced at the other closed end to take a double strand of twisted brass wire. The Italian, Guiseppe Console, invented this ignitor in 1835 for use in the percussion lock adopted by the Austrian government for converting flintlock muskets to percussion.

Maynard tape primer formed of pellets of priming composition cemented between two strips of shellacked paper. The larger rolls, brown in color, shown at top were wrapped in brown paper and packed in Japanned tins of ten rolls. Each roll contained about 40 primers. The smaller roll with its tin cover shown below is red, and contains only 36 primers. A blue variant exists. The tin cover is embossed: "Maynard's Patent, Sep. 22, 1845. Manufactured by MASS ARMS Co., Chicopee Falls, Mass."

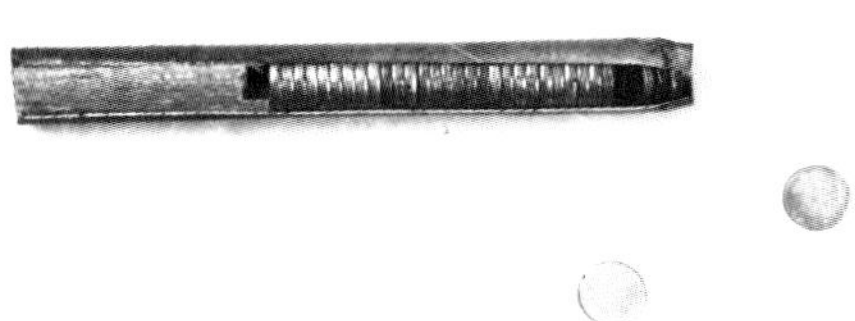

Sharps disc primer charger—a brass tube open on one side containing 26 copper discs formed of two pieces with fulminate between them, the remaining length of the tube filled with a wood spacer which serves to push the discs into the primer magazine. The number of primers varies from tube to tube.

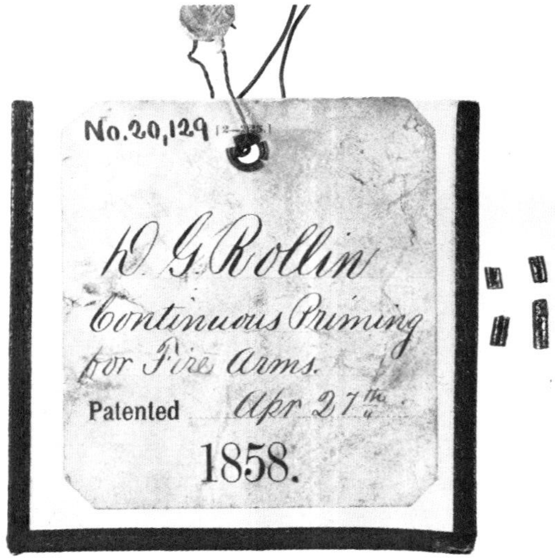

Description

Rollin's continuous priming in the form of a small cylinder of fulminate composition which was fed from a magazine to be struck and exploded by the hammer. This evidently required a different form of nipple leading into the chamber and a hammer nose which would descend into a recess designed to contain this stick of fulminate.

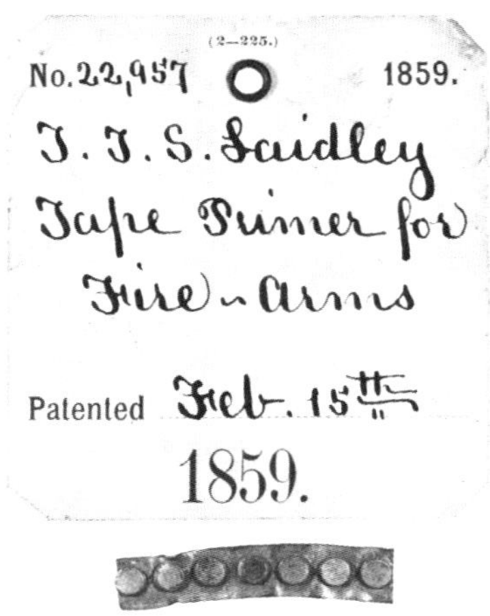

The Laidley tape primer was formed out of a strip of lead with pockets in which priming composition was deposited. Hammer and lock action worked the tape out of a magazine to place one priming-filled pocket at a time over the percussion nipple. When the hammer nose struck the primer, the tape was cut off at the same time, presumably preventing ignition of the remaining primers. Apparently, the Rollin and Laidley priming systems were never put into general production and use.

Tube, Tape, and Disc Primed Locks

Description

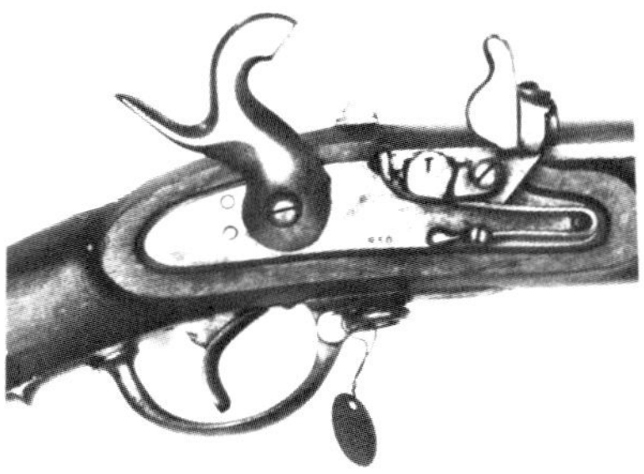

Lock replacing the flint type with a cover hinged at the front of the pan. The tube ignitor was inserted into the touchhole on the side of the chamber and the cover closed over it. The hammer nose struck the top of the pan cover, crushing and exploding that end of the tube projecting from the touchhole, thus igniting the charge. The specimen shown is that improvement of the original Console lock developed by Austrian Major General Augustin.

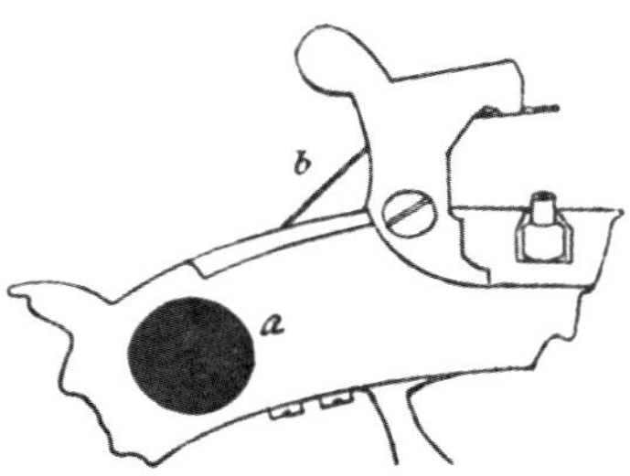

Magazine in the small of the rifle stock containing a roll of priming pellets between thin strips of copper. A pawl in the hammer itself fed the strip forward to position a pellet on the nose of the hammer. This system was used on the Sardinian Bersaglieri rifle, 1844.

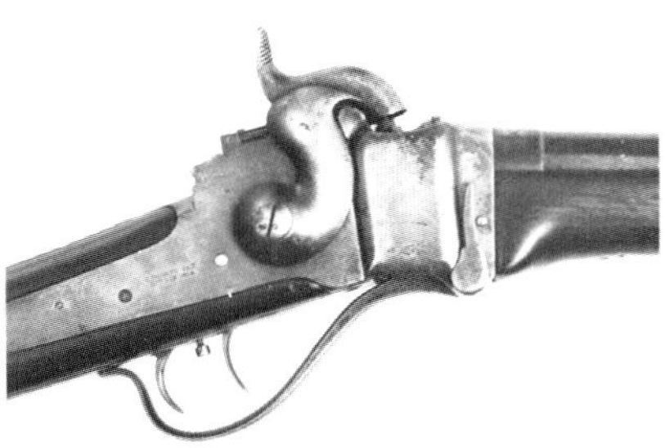

The Sharps disc priming device was a magazine located on top of the lock plate beneath the hammer in which a column of disc primers was held by spring action. Hammer action operated a mechanism which threw one disc at a time out over the percussion nipple just as the hammer fell to crush and explode it.

(Drawing from "Geschichte der Handfeuer-waffen," Dr. Reinhold Gunther, 1909. Sharps rifle collection of Steve Brenger.)

Locks Fitted With Tape and Disc Priming Systems

Description

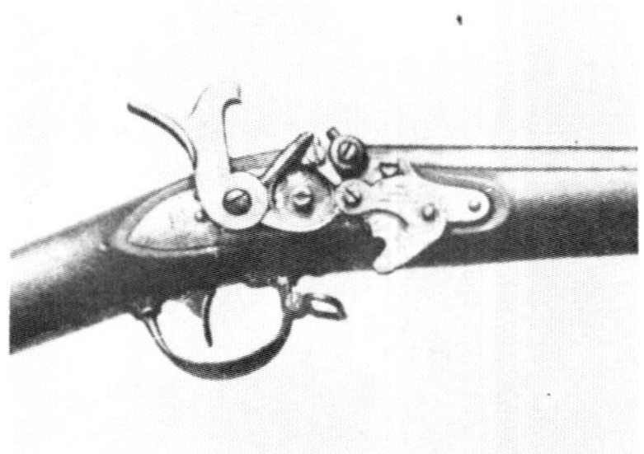

Smoothbore percussion musket converted from flintlock and marked "D. Nippes, Mill Creek, Pa. 1843." Fitted with an early Maynard tape primer magazine on the lockplate. The magazine cover is shown open.

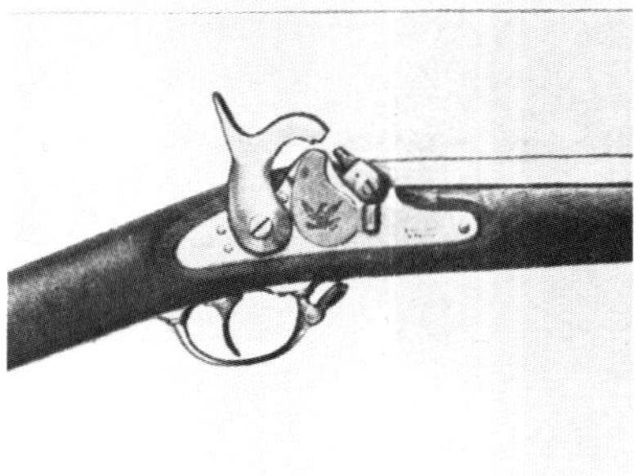

Model 1855 U.S. rifled musket marked "E. Whitney N. Haven 1858." The Maynard tape primer magazine is shown closed.

Percussion musket converted from flintlock and rifled, with tape priming magazine manufactured by Remington, Ilion, N.Y. This magazine differs slightly from the Maynard in the springs which advance the tape primers onto the cone. It also has a small knob to aid in opening the cover of the magazine. This type was made under Maynard's patent.

Originally a flintlock musket, altered to be percussion in 1858. Lock marked "L. Pomeroy, 1823, U.S." The tape priming system contained in the hammer was patented by J. N. Ward on July 1, 1856.

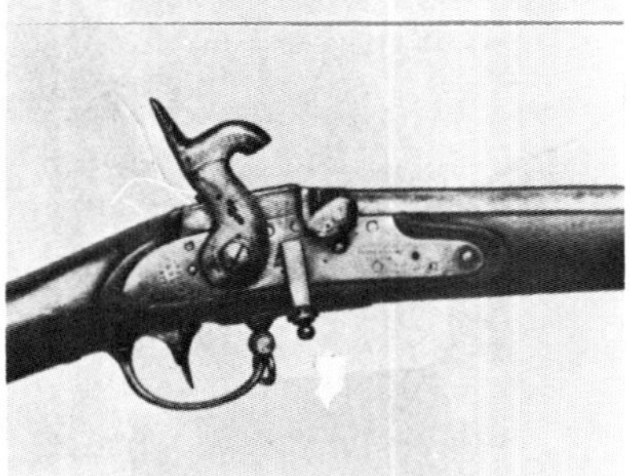

United States flintlock musket, 1824, altered to percussion. Below the lock plate is a cylinder containing disc primers. The mechanism fed the discs to the nipple by action of the hammer.

(Illustrations from an original 1928 catalog of the Nunnemacher Arms Collection, courtesy Milwaukee Public Museum)

4 THE PERCUSSION SMOOTHBORE MUZZLE-LOADER

Despite the introduction of the rifle during the late flintlock period, the smoothbore musket continued to be the primary infantry long firearm well after the acceptance and gradual adoption of percussion ignition. Smoothbores dictated the use of round lead projectiles, or buck and ball and buckshot auxiliary loads. Consequently, the cartridge for the percussion smoothbore musket differs little in general appearance from the cartridge for the flintlock smoothbore.

The main difference to be noted is the reduced length of the case allowed by a smaller powder charge. This occurred as a result of improvements in gunpowder manufacture.

Advancements in Black Powder Manufacture

By 1340, the Germans were making what came to be known as "corn" powder. A quantity of meal powder was placed in a receptacle and moistened with water, urine, or vinegar. This formed cakes after drying which were then pulverized. The fragments were forced through small holes in a stretched hide and further passed over a horsehair sieve to separate the fine dust from the granules. The appearance of these granules led to their name. The dust powder was used for priming.

This corn powder was far more stable than the early ground up variety because the grains held the components together. The ingredients of black powder made in the American colonies before and during the Revolutionary war was 75.2 percent saltpetre; 11.3 of sulphur; 13.5 of charcoal.

Black powder continued to be perfected during the first part of the 19th century. Some forms in Europe were of irregular oval shape, those in America of irregular shape which were classified according to grain size. This improved powder required only one-half to two-thirds the charge used during the 18th century to produce the same propelling energy.

References: "Small Arms and Ammunition in the United States Service, Col. Berkeley R. Lewis, 1956.

PSB 1 – .753in (19.13mm) English Musket, Model 1842

Description

Tan paper tube tied over a round ball and secured by a second string tie between ball and powder charge. Tail twisted closed and flattened along the cartridge. The ball is exposed at the nose of this specimen.

PSB 2 – 17.5mm (.69in) Swiss Musket, Model 1842

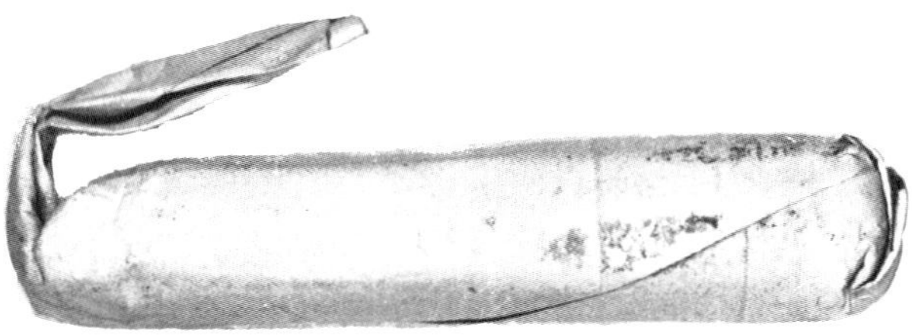

White paper tube folded and cemented over a round ball. Tail flattened, folded, from each side to center, and laid along the case.

PSB 3 – .69in (17.5mm) U.S. Musket, Model 1841, Buck and Ball

Light brown paper tube tied over three .31in shot loaded atop a single full-size ball. Tail folded from each side to center and laid along the case.

Comments

The English commenced work on a percussion musket in 1834 but did not issue such arms generally until 1842. PSB 1 above was used in the early .753in Brown Bess flintlock muskets converted to percussion as well as the new percussion Sea Service and infantry muskets. The cartridge for the percussion models had a smaller powder charge than that used in the earlier flintlock since better powder was then available and the 12 to 15 grains formerly used to prime the flintlock from the cartridge itself was no longer needed.

The Swiss cartridge above came from a packet labeled "Zeughaus Zurich." Others made in Canton Bern and Canton Unterwalden, Switzerland contain cartridges about two-thirds the length of this specimen and may be for the cadet musket.

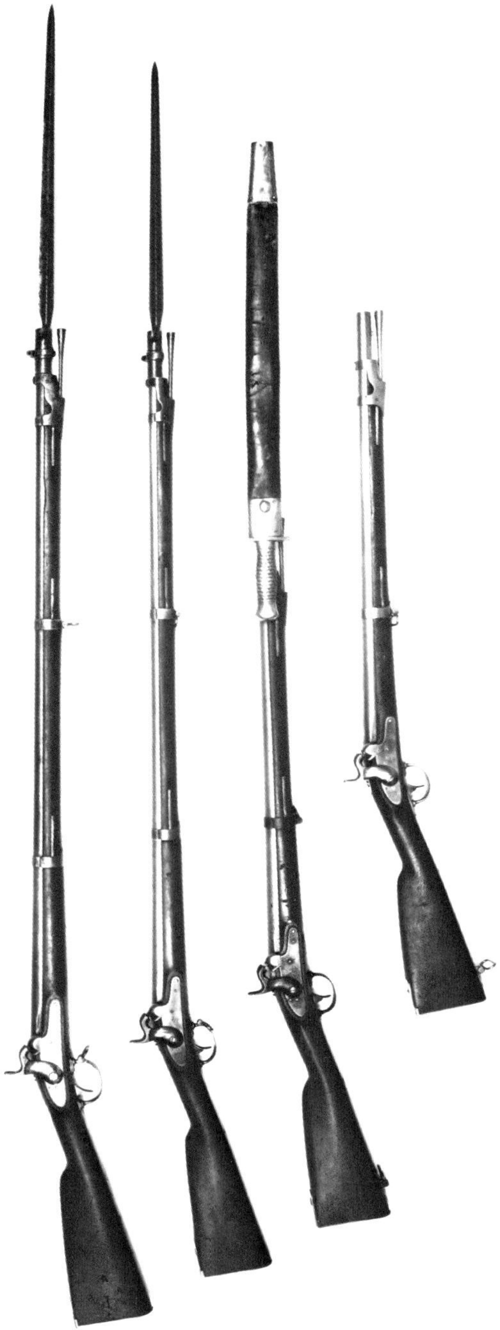

Group of United States smoothbore long arms in the Nunnemacher Collection.

From left: Model 1842 musket, caliber .69, lock marked "Springfield 1844;" .58 caliber musket, lock marked "Springfield 1853;" Model 1847 sapper's musketoon with sword bayonet; Model 1847 artillery musketoon.

These were the last of the U.S. smoothbore arms and differed from those used by other nations only in minor details, furniture etc. At the outset of the American Civil War, 1861, the antagonists were armed in large part with the smoothbore musket and tens of thousands were imported from European countries by the Union and the Confederacy. Ultimately, many of the domestic arms were rifled and equipped with rear sights to fire the conical "Minie" bullet (see Chapter 5). Following the war, some quantity of these rifles were then transformed by various breech-loading systems during the early experimentation with self-contained metallic ammunition (see chapters 5, 6, and 10. Milwaukee Public Museum photo).

PSB 4 – 17.8mm (.70in) Austrian Console Musket, 1837-1842

Description

White paper tube cemented closed over a lead ball surrounded by two felt bore-size wads. The powder column is closed with a cardboard disc and twist of the tail. A flattened "Zunder," or percussion tube ignitor, is attached to the base disc by a double strand of twisted brass wire.

PSB 5 – 17.5mm (.69in) Dutch Musket, ca. 1840

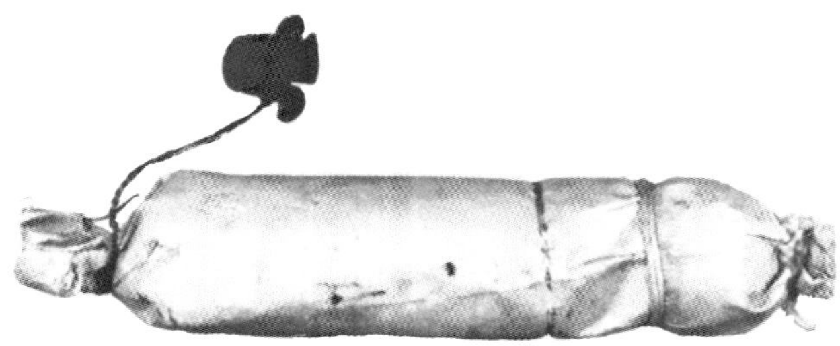

Tan paper tube string tied over and below the ball. Powder column is closed with a twisted length of brass wire to the loose end of which is attached a percussion cap.

Comments

These specimens illustrate early attempts to incorporate the priming device with the cartridge itself. Giuseppe Console, an Italian, designed a percussion lock in 1835 which employed a tube ignitor made of rolled brass foil. The Austrians adopted this system to convert flintlock muzzle-loaders to percussion in 1837. By 1842, Austrian Major General Vincenc Augustin had improved the lock and the ignitor was flattened to the form shown attached to PSB 4 above. In loading the Console lock, the ignitor was inserted in the touch-hole and the cover was brought down over it. Then the soldier pulled the base closure wad out of the cartridge by the wire to open the powder column. The powder was poured down the barrel and the entire wad and ball unit rammed down on the powder charge. This system obviated the necessity of biting open the cartridge, and unified cartridge and priming device.

The Dutch cartridge had to be opened by hand to expose the powder for loading. Its only advantage was in attaching the primer to the case. It would appear, however, that the clipped-off tail required the case to be opened by unwinding the wire because tearing it open would spill a quantity of the powder charge.

PSB 6 – 16.5mm (.65in) Danish Voss Rocket Projectile, 1834

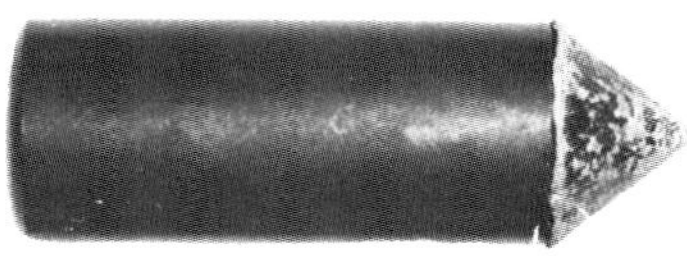

Description

Drawn brass cylinder turned over at the base leaving a central ignition orifice. Mouth of the cylinder is filled with a conical lead projectile and the remainder of the cylinder is filled with a combustible mixture.

PSB 7 – 17.52mm (.69in) Bavarian Incendiary Rocket, ca. 1840

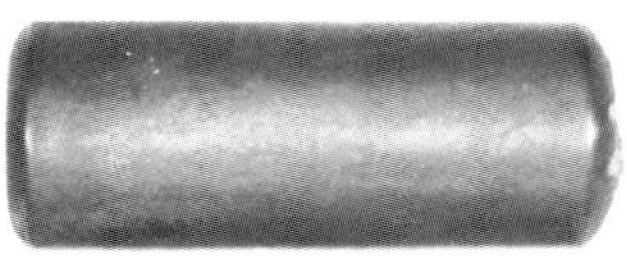

Solid drawn brass cylinder turned over at the nose and base with a lead cylindrical filler occupying one-fourth of the forward interior, the remaining cavity filled with incendiary mixture. Ignition orifice located in the center of the base. The pale green wrapper in which this specimen was found is marked in handwritten ink– "Brandraket fur Musketen" (incendiary rocket for muskets).

Comments

Chief of Danish Ordnance Voss developed the first specimen shown. These projectiles were loaded atop a black powder charge in a smoothbore musket. The exploding powder ignited the combustible incendiary mixture within the cylinder's rear section through the base orifice. Moritiz Thierbach describes and illustrates these in "Die Geschichtliche Entwicklung der Handfeuerwaffen." 1886-1899, (History of the Development of Small Arms).

(Note: Voss developed cartridges similar to PSB 6 above during his career which were used in breech-loading rifles. Others he designed contained their own priming. They superficially resembled the specimen shown above.)

PSB 8 – 17.5mm (.689in) Danish Espingol Cartridge for 20-barrel "Orgelespingol"

Description

Cardboard cylinder closed at the base with a disc of cartridge paper, filled with 5 grams (77.1 grains) of musket powder and completed with a conical bullet with wide belt. The bullet has a hole running through it from nose to base.

PSB 9 – .64in (16.26mm) U.S. Colt Revolving Musket and Rifle-Musket, Mod. 1855

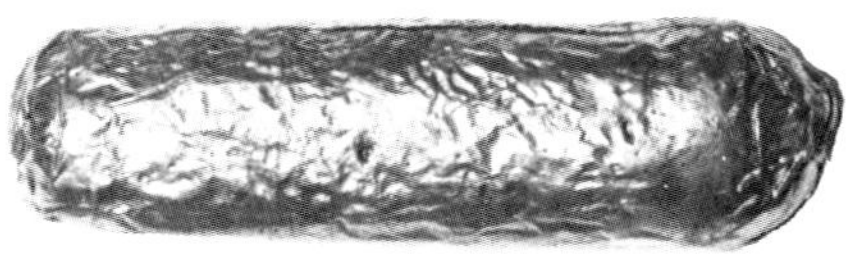

Case made of tinfoil closed over a round ball with powder charge filled in on top of a round ball. The tail is folded and pressed to close the powder column.

Comments

The Danish Espingol cartridge was developed by inventor N. Löbnitz in the late 1840's. As described opposite, 10 to 15 of these cartridges were loaded one after the other into musket barrels mounted together. Although for smoothbore barrel, the bullet was belted to keep the hole in the bullet aligned from front to rear. The percussion lock mounted near the muzzle of the barrel fired the cartridge closest to the muzzle first and the powder explosion from that round flashed back through the hole in the bullet directly behind it, transmitting the flame to the next powder charge. The cartridges were ignited in this way one after the other until all were fired in Roman candle style.

Tinfoil case cartridges, although very rare, have been known in handgun sizes for many years. However, the specimen shown at PSB 9 is one of only ten known to have survived in a packet presumably from 1844. Col. Lewis reports from the records of the Chief of Ordnance for the year ending June 30, 1844 that 150,000 tinfoil musket cartridges and 50,000 each for rifle and carbine were purchased that year by the U.S. military. The label on this packet in a private collection reads— "Colts Tinfoil Cartridge Musket." Evidently Colt is the only known manufacturer of this ammunition in the U.S., though English patent 1324 of June 9, 1855 covers the same idea. The tinfoil cartridge was whole inserted, the percussion flame penetrating the case to ignite the powder.

Reference: "Small Arms and Ammunition in the United States Service," Berkeley R. Lewis, 1956. (Illustration courtesy the Danish National Museum, Copenhagen, Denmark from a 1949 edition of the "Tøjhusmuseet Skrifter")

S. Löbnitz's Orgelespingol, Model 1864, in the Heeresmuseum, Zeughaus, Berlin, Germany. Ten musket barrels per row were attached to a common breech plate and to iron bars situated on top and below the barrels connected to a bolt on either side to hold the barrels rigid. The percussion locks are shown mounted near the muzzles of the barrels rather than at the breech in an ordinary muzzle-loading firearm.

To operate, each barrel was loaded with 10 to 15 cartridges one after the other. Only one percussion cap per barrel was needed to set off the entire train, starting with the cartridge nearest the muzzle. Presumable, all of the barrels could be fired at one time, since the description indicates that up to 300 shots could be fired in a minute.

This was an advancement of an idea which first appeared in the 14th century. Multiple barrel weapons called "organ" guns were constructed by mounting a number of musket barrels on a frame or wheeled carriage. These were found useful to protect foot troops and artillery units from cavalry charges. The idea survives today in the multiple fire rocket weapons used by a number of nations in ground units and aboard naval vessels (Illustration from same source as PSB 8, courtesy the Danish National Museum)

PERCUSSION SMOOTHBORE ARMS AND CARTRIDGE DATA

No.	Bore	Model and Date	Ball Diameter	Ball Weight	Powder Charge	Cartridge Length	Sources
PSB 1	.753in 19.13mm	English Musket, Model 1842	.692in 15.98mm			2.64in 67.06mm	Specimen
PSB 2	17.7mm .69in	Swiss, Model 1842	.670in 17.02mm				
PSB 3	.69in 17.5mm	United States, Model 1841	.667in 16.94mm	505grs 32.72gm	130grs 8.42gm	2.55in 64.77mm	Specimen—Lewis
PSB 4	17.8mm .70in	Austrian Console, Model 1837-1842	17.27mm .680in	22gm 339.5grs	4.2gm 64.8grs	67.5mm 2.66in	Specimen—Dolleczek
PSB 5	17.5mm .69in	Dutch, ca 1840	17.73mm .698in			87mm 3.43in	Specimen
PSB 6	16.5mm .65in	Danish Voss, Model 1834	16.81mm .662in			49.28mm 1.94in	Specimen
PSB 7	17.52mm .69in	Bavarian, ca 1840	15.5mm 61mm			40.5mm 1.59in	Specimen
PSB 8	17.5mm .689in	Danish Espingol, ca 1849-1864	16.70mm .657in			49mm 1.93in	Tojhusmuseet, Copenhagen
PSB 9	.58in 14.73mm	U.S. Musket, 1844	.632in 16.05mm			2.16in 54.86mm	Specimen—Lewis

Note: The projectile diameters given above, except for PSB 8, were taken over the paper case or the metal cylinder as in PSB 6 and 7. The actual ball diameter in the paper cartridges will be from .10 to .20 thousandths smaller, depending on the thickness of the paper.

5 FROM ROUND BALL TO "MINIE" BULLET

The rifle remained a secondary firearm in the world's armies until the early 1840's. Pounding a bore-fitting ball down the bore with a strout ramrod slowed the rate of fire and exhausted the soldier to the extent that he could not do accurate shooting.

The French undertook experiments in 1828 which ultimately solved this problem. Captain M. Delvigne, an officer in the Royal Guard, devised a chamber which was smaller than the interior diameter of the barrel. The chamber formed a shoulder upon which the ball could rest. After loading the powder, the shooter dropped a ball smaller than the bore down the barrel and then gave it two or three sharp raps against the shoulder with his ramrod to upset and spread the ball into the rifling.

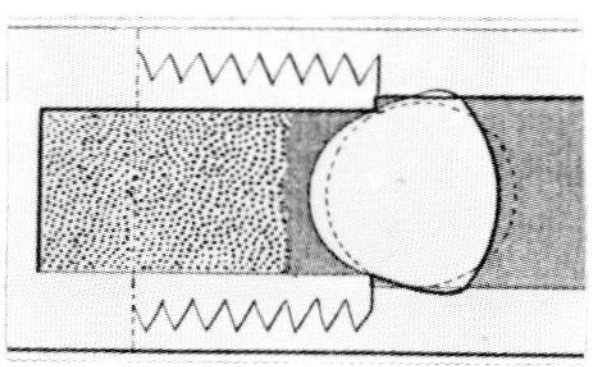

Capt. Delvigne's chamber

Delvigne's system did what he expected. It made loading a simple, effortless process, eliminated the windage, and permitted the rifling to spin the ball. Although he cupped the head of the ramrod to fit the ball, the projectile was deformed by the blows it received against the chamber shoulder. Its out-of-round shape caused it to fly erractically with adverse effect on accuracy. In additon, fine shavings or bits of lead were cut from the ball as it struck against the sharp edge of the shoulder. These fell into the bottom of the chamber where they gradually built up and reduced the powder capacity of the chamber.

Colonel Pontchara's Improvement

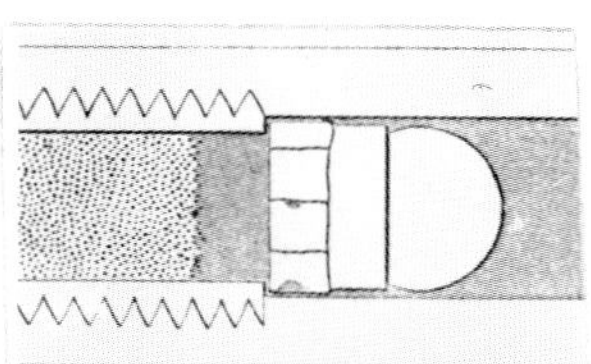

Col. Pontchara's Improvement

Colonel Pontchara, another French officer, modified this system by adding a wooden cylindrical sabot cupped out to receive the ball and covered at its base end with a greased fabric patch. The sabot was intended to eliminate the cutting and disfiguring of the ball against the shoulder, and the greased patch was intended to ease loading and clean the bore. This was an improvement but the wooden sabot tended to break up under blows from the ramrod to expand the ball and it did nothing to lessen the disfiguring of the ball caused by flattening it into the rifling. Nonetheless, ten battalions of French troops were armed with this Delvigne-Pontchara chambered rifle in 1840 and the Belgians adopted a similar system about 1842.

The Austrian Improvement

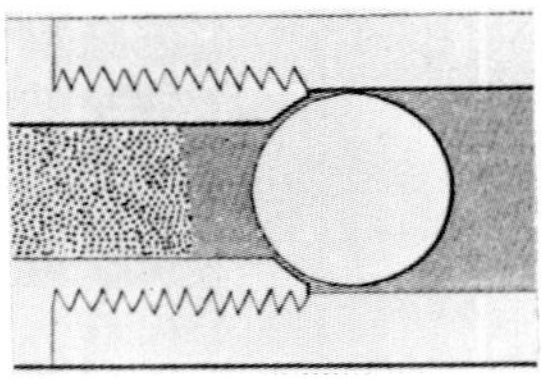

Gen. Augustin's Chamber

The Austrian Major General Vincenc Augustin, observing the French accomplishments, decided to adopt the same idea but he dispensed with the wooden sabot and modified Delvigne's chamber by grinding away its sharp shoulder at a 45 degree angle to fit the sides of the ball. This eliminated the shaving action of the chamber shoulder but the ball was still disfigured when struck by the cupped ramrod to expand it into the grooves of the barrel.

Meanwhile, Captain Delvigne had already experimented with an elongated cylindro-conical form of bullet, and it is he who is commonly credited with discovering that a bullet hollowed out at its base was expanded by the exploding powder gases into the grooves and lands of the rifled barrel. However, the French military authorities were not ready to accept the idea of an elongated bullet to replace the long-established round ball. Captain Norton, a British officer, designed a similar elongated, hollow-based bullet in 1823 which contained the powder charge within its base. However, he got nowhere with the Select Committee on Firearms who dismissed it as a novelty. To the Committee, a spherical ball was the only shape of projectile adapted for military purposes.

William Greener devised an oval bullet in 1835 with tapered hollow in its base into which he fitted a pewter plug with a head like a round-topped button. The wedge functioned to expand the bullet into the walls of the barrel when the powder charge drove it forward into the bullet's hollow base. This idea met with similar rejection by British military authorities, although Greener was later rewarded on demand for having first developed this expanding bullet idea in England.

Thouvenin's Experiments

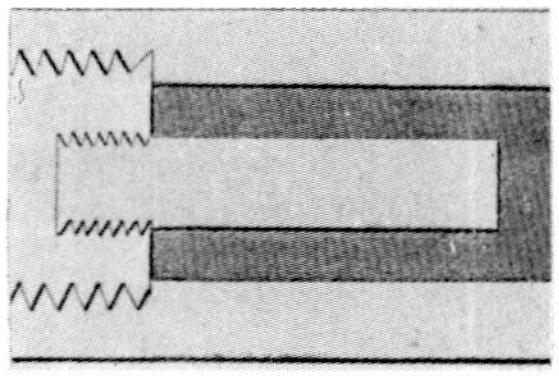

Col. Thouvenin's "Tige" chamber

Next in line of progression came the experiments of French Artillery Colonel Thouvenin, who substituted for the shouldered chamber a steel pillar, or "tige," half the diameter of the bore which rose from the base of the chamber. The powder charge dropped in around this pillar which was long enough so that the ball rested against it without compressing the charge. Colonel Thouvenin first experimented with the round ball but was unsuccessful in improving the rifle's performance. Therefore, he decided to adopt Delvigne's cylindro-conical bullet with solid base and single groove around its waist. His assistant, one

Captain Tamisier, suggested a number of smaller grooves around the bullet's cylindrical portion and constructing it so the center of gravity was nearer to the point. This change was made. A ramrod with head hollowed out to fit the nose of the bullet was used to expand the bullet's base into the rifling.

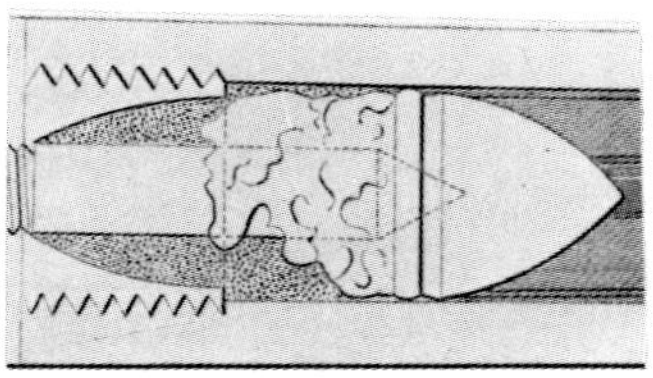

Chamber of the Oldenburg tige rifle

This idea worked far better than its predecessors. Although it was difficult to clean this rifle around the tige, and this projection into the chamber was subject to bending by heavy blows of the ramrod, the French adopted it in 1846. Chambered rifles were altered to tige rifles and several other countries soon followed the French lead by adopting this system with minor modifications of their own.

The "Minie" Bullet

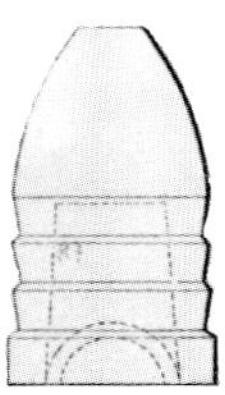

Nassau Minie bullet with culot

It was not long, however, before Captain Claude-Etienne Minie, Inspector of Musketry at the Military School of Vincennes, developed a cylindro-ogival bullet with three angular grooves filled with lubricating tallow and a cavity conical from the base upwards. He closed this cavity with a sheet-iron cup or "culot." The function of the culot was to serve as a wedge to expand the bullet into the rifling from the force of the exploding powder charge.

This system required no pillar and no chamber smaller than the bore. The bullet was not rammed, but pressed home against the powder, so it suffered no deformation during loading. The first Minie bullet worked well, but the culot sometimes loosened during loading or firing and flew off at an angle after leaving the bore.

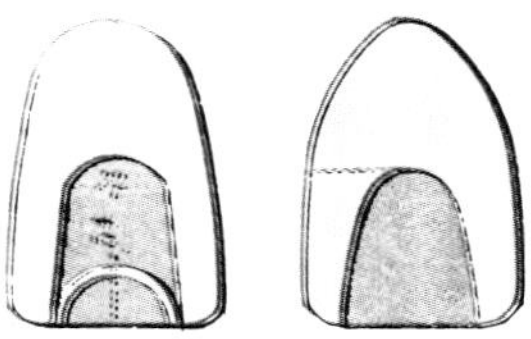

English Minie bullets with and without the culot

Again, this system was adopted with modifications throughout Europe. However, the culot was ultimately found to be unnecessary and was abandoned. All that was necessary to ensure expansion of the hollow-based bullet was to construct the cavity properly so that the skirt would spread into the rifling and there was sufficient thickness of lead at the nose to prevent the exploding powder from forcing its way through the bullet altogether.

Wilkinson-Lorenz and Swiss Bullets

The French were by no means alone in their attempts to perfect the muzzle-loading rifle. Wilkinson in England and Ritter von Lorenz in Austria independently developed bullets with deep grooves which compressed from the force of the exploding powder charge to spread the bullet into the rifling. The English tried this expansion bullet in the Enfield rifle but chose the Minie type instead. The Austrians adopted and used it in their rifles for several years.

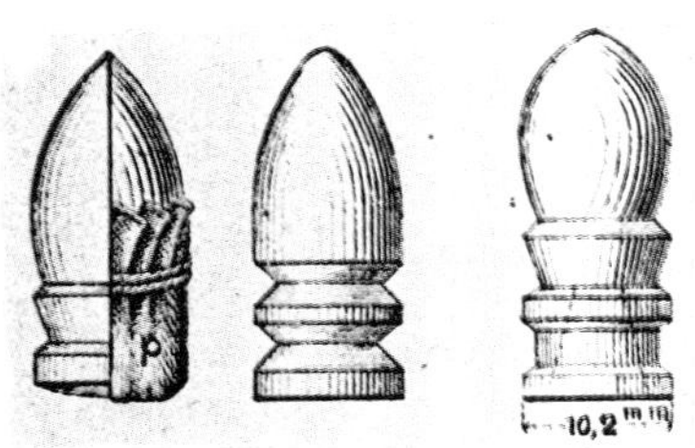

The Swiss, after experiments with all of these types, settled on a shallow-grooved semi-hollow based bullet in the remarkably small caliber of 10.4mm (.409in) which made their Federal rifle adopted in 1850 the longest range, most accurate military rifle of the day. This Swiss bullet, long for its weight and therefore, ballistically efficient, pointed the way for reductions in bore size.

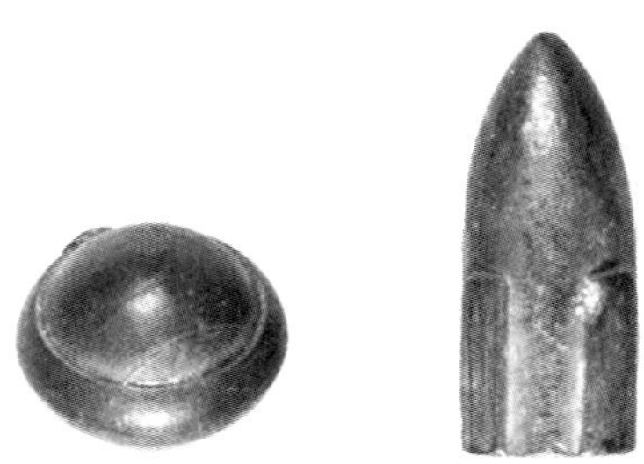

Belted ball and four-groove bullet used in English Brunswick two-groove and Jacob four-groove rifles. Powder and bullet were loaded separately.

6 THE PERCUSSION RIFLED MUZZLE-LOADER

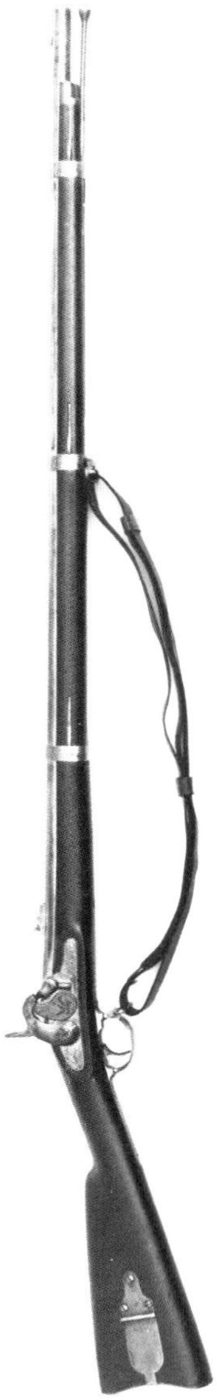

U.S. Rifle-Musket, Model 1855

The era of the percussion rifled muzzle-loader extended from about 1841 to 1865. This broadly overlapped the needlefire and percussion breech-loaders. During this time, the round ball gave way to the elongated cylindro-conical bullet as the primary small arms projectile.

Typical of these arms is the U.S. percussion rifle-musket, Model 1855, caliber .58 (14.73mm). Made at Springfield Arsenal in 1860, it is fitted with the Maynard tape priming magazine. Since the rifle-musket used during that era differs from nation to nation only in minor details (style of stock, furniture, shape of hammer, lock, etc.,) the specimen shown at left may be considered representative of the arms that fired the various cartridges shown in this chapter, with certain exceptions illustrated because of their unusual characteristics.

To load the Model 1855 rifle-musket, the soldier tore the cartridge open at the tail, usually with his teeth. He poured the powder charge down the barrel and then gripped the ball and struck the empty paper case at the base of the bullet across the muzzle of the rifle to break the paper there. This exposed the bullet's base and allowed it to be forced base first into the muzzle. The paper covering over the bullet was discarded and the ramrod used to push the projectile down against the powder charge, taking care not to crush the powder. (Rifle in Steve Brenger collection)

Label from 10-round packet of ammunition for .58 cal. rifle made at St. Louis Arsenal, 1864.

Description

PR 1 – .54in (13.72mm) U.S. Harpers Ferry Rifle, Model 1841

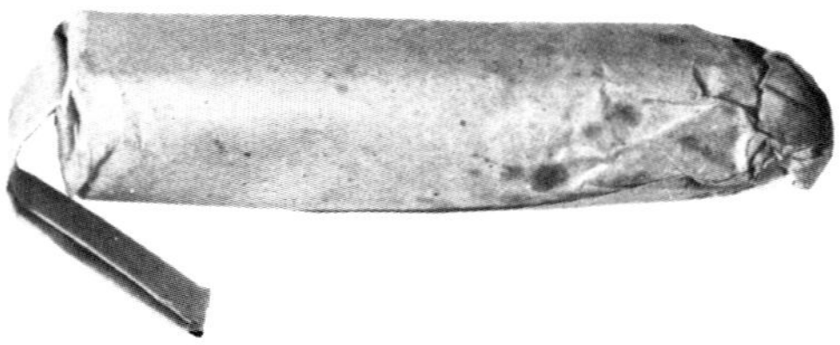

Brown paper case tied closed over a conical, grooved, hollow-base "Minie" bullet. Tail flattened to close the powder column, folded from sides to center, and laid along the case.

PR 2 – .69in (17.5mm) U.S. Rifle-Musket, Model 1842 (altered)

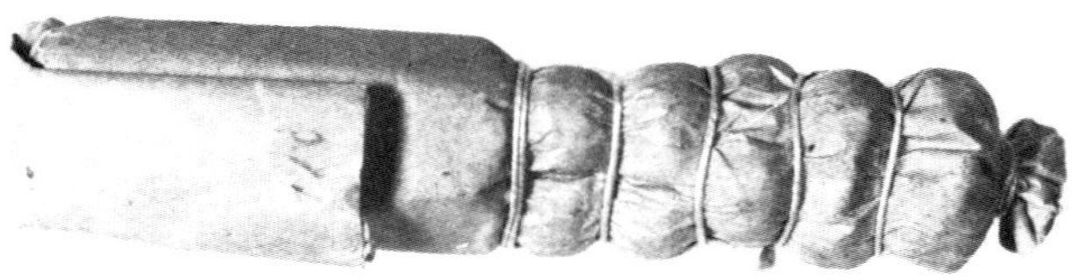

Brown paper tube tied over, between, and below five tiers of three each .31in buckshot. Tail folded as above.

PR 3 – .58in (14.73mm) U.S. Rifle-Musket, Rifle Model 1841 (altered from smoothbore,) and Model 1855 rifle

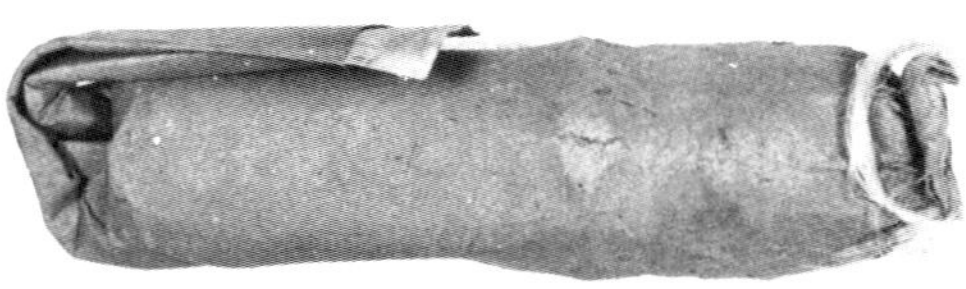

Salmon paper case with separate inner powder container of white paper folded in with tail of case and showing at the tip. Otherwise constructed as PR 1. This cartridge contains the Williams clean-out bullet with zinc washer attached to its base.

Comments

The cartridge first issued for the Harpers Ferry rifle of 1841 was loaded with round ball. The conical bullet load was introduced later and a nearly identical round was used by Confederate forces during the American Civil War.

The .69 buckshot load was an auxiliary cartridge for the .69 cal. smoothbore altered to rifle for a conical projectile. Buck and ball loads were also used.

The U.S. Minie bullet was lubricated to fill the grooves with one part beeswax and three parts tallow, applied hot before making up the cartridge. The specimen shown at PR 3 was usually included one or two to a ten-round packet of ammunition. The zinc washer served to scrub fouling from the rifle bore. The Williams clean-out cartridge will also be found with blue dyed paper envelope, or the common tan cartridge case.

PR 4 – .54 Rifle, Confederate States of America, manufactured at Fayetteville, North Carolina arsenal, 1861-1865.

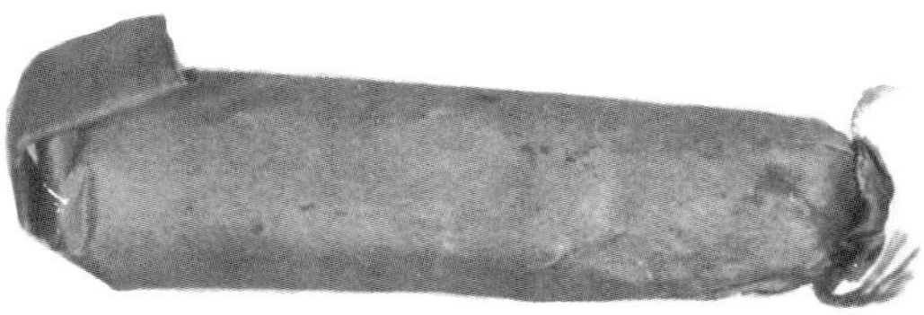

Description

Salmon-color paper tube with conical ball made in the same fashion as the U.S. PR 1 and PR 3.

PR 5 – .69in (17.5mm) Rifled Musket, Confederate States of America, ca. 1861-1865

Pink or white paper tube swaged into the base of a two-groove, hollow-base minie bullet. The filled powder column is closed by flattened tail twisted and laid along the case.

Comments

Confederate arsenals followed the pattern of U.S.-made ammunition except for the distinctive inserted paper type shown at PR 5. These cartridges are found in at least .69, .64, .577, .54, and .45 calibers, the last for an unknown arm.

Description

PR 6 – 18mm (.708in) French Rifle, Model 1842

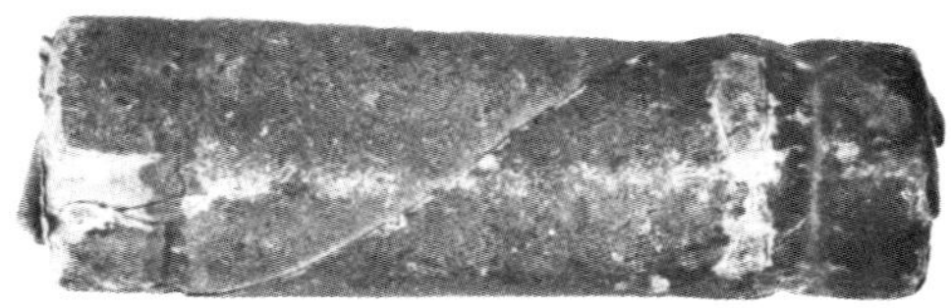

Brown paper tube with "Minie" bullet loaded in rear of tube with powder charge filled in over nose of the bullet. End of tube folded into the hollow base of bullet which is lubricated. Powder column closed by twist of the tail which is pushed into the cavity above the powder charge.

PR 7 – .69in (17.53mm) Mexican Rifled Musket, ca 1850

Gray-green unsized paper case containing a hollow-based Minie bullet with powder charge filled over the bullet. Paper case folded into the hollow of the lubricated bullet and the tail flattened, folded from both sides to the center, and laid along the case.

PR 8 – 17.5mm (.689in) Belgian Minie Rifle, ca 1863

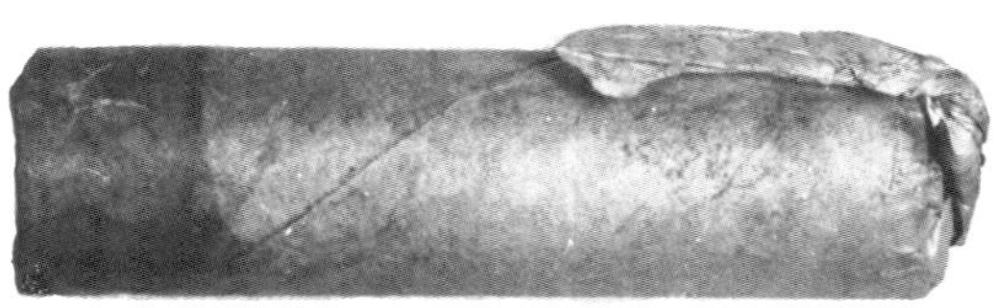

Tan paper tube with outer wrap of very thin paper, both pressed into the cavity of a lubricated Minie bullet. Projectile is the Timmerhans type with bell-shaped cavity in the base. Tail twisted to close the powder column and laid along the case.

Comments

These three specimens typify the cartridge that was developed for the Minie rifle, the Mexican army adopting the European pattern. Some were made so that the bullet was separated from the powder by a wad of paper formed around the bullet nose, or the powder was contained in a separate inner tube of thin paper.

PR 9 – .702in (17.83mm) British Minie Rifle, Model 1851

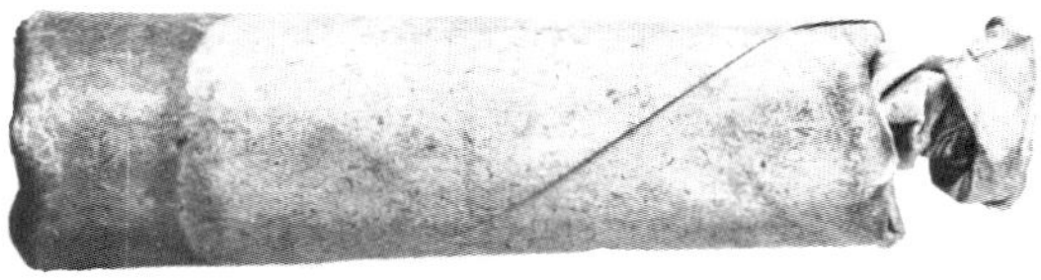

Description

Brown paper tube loaded with a single-groove conical bullet with powder charge filled in over the nose of the bullet. Paper pressed into the hollow base of the bullet and twisted closed over the powder.

PR 10(a) – .577in (14.65mm) British Enfield-Pritchett Rifle, Model 1853

Off-white paper tube loaded with a lubricated, grooveless, conical lead bullet with bell-shaped hollow base. Powder charged filled in over the bullet. Paper is pressed into hollow base of bullet and twisted closed over the powder.

PR 10(b) – .577in (14.65mm) British Enfield, Model 1853

Later form built up from an inner stiffening tube within a paper powder wrapper, green in this specimen, one end of which is folded over the powder into the stiffener's base to form a cavity. A lubricated, grooveless, conical lead bullet with boxwood or clay plug in its hollow base is fitted nose first into this cavity. An outer wrapping of white paper joins the powder charge to the bullet. It is folded against the bullet's base and tied. The other end of the powder wrapper is twisted closed to form the tail.

Comments

In the early 1850's the British started converting over from smoothbores and rifled the 1842 pattern musket of .75 caliber with three or four grooves to a bore diameter of .758in (19.25mm). It required a cartridge even larger in diameter than PR 9 shown above. Efforts begun in 1852 at the Royal Armory, Enfield, resulted in the .577in rifle which replaced the early Minie rifles.

In loading PR 10(b), the tail was torn off, the powder poured down the barrel, and the whole remaining cartridge reversed. The bullet end was inserted as far as lubricant extended before the powder tube and stiffener were knocked or torn off. Then the bullet was pressed home on the powder with the ramrod.

PR 11 – 14.65mm (.577in) Turkish Enfield, ca 1860

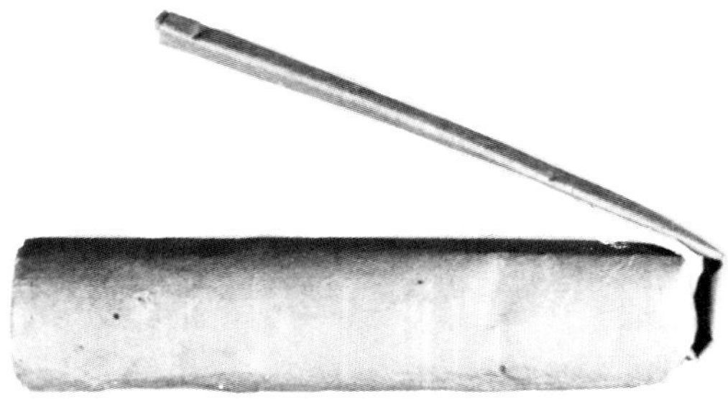

Description

Tan paper tube folded into the base of a Minie-type bullet. Tail double folded to center and laid along the case. Bullet has a wood plug in its base.

PR 12 – 15.24mm (.600in) Russian "6-Line" Rifle, Model 1857

Grey paper tube folded into a hollow-base grooved bullet and twisted closed over the powder column. Tail snipped short.

Comments

The Turkish cartridge shown was used in the same type rifle as the British Enfield. However, it exhibits none of the sophisticated features of the British ammunition.

The Russian cartridge is the ultimate in simplicity. A pinch of the powder column at its upper end opens the short, twisted tail sufficiently to expose the powder. It seems questionable, however, that a cartridge with such a short tail could withstand much handling without the powder charge escaping little by little from the case. The Russian term "Line" was an abbreviated system denoting 2.54mm (approximately .10in). The term carried over into the smokeless powder era when the caliber of the infantry rifle was reduced to 7.62mm (.30in), just one-half of the 15.24mm rifle of the late muzzle-loading era. Headstamps on the 7.62mm Russian cartridges often contain the symbol ''' at 6 o'clock–3 lines.

Description

PR 13 – 13.9mm (.547in) Austrian Rifled Musket, Model 1855

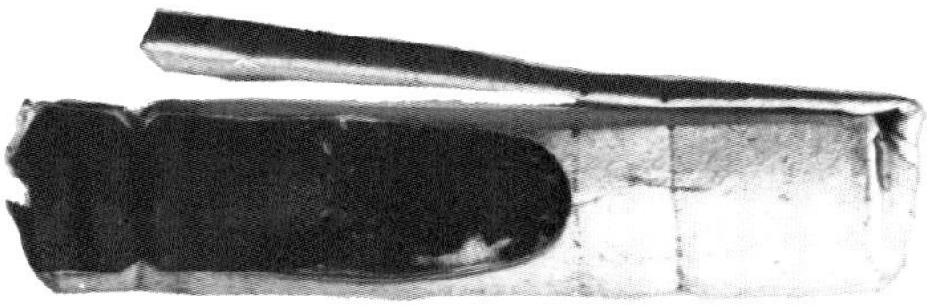

White paper tube cemented over base of a two-groove "Lorenz" compression bullet. Tail double folded to center and laid along case.

PR 14 – 13.9mm (.547in) Prussian Rifle, Model 1864

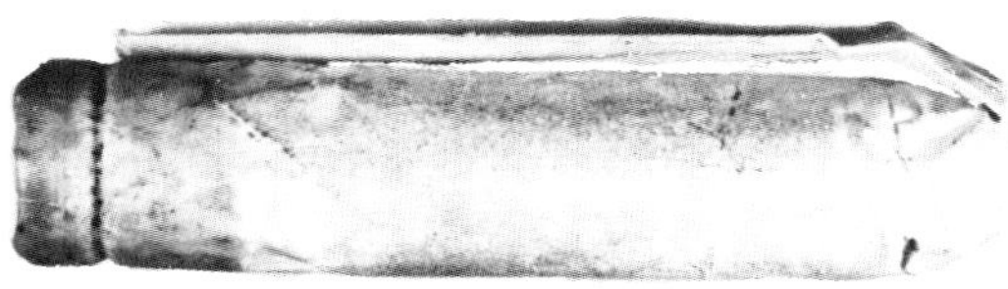

White paper case constructed essentially as above with same type bullet which is lubricated.

PR 15 – 12.5mm (.492in) Dutch Rifle, Model 1863

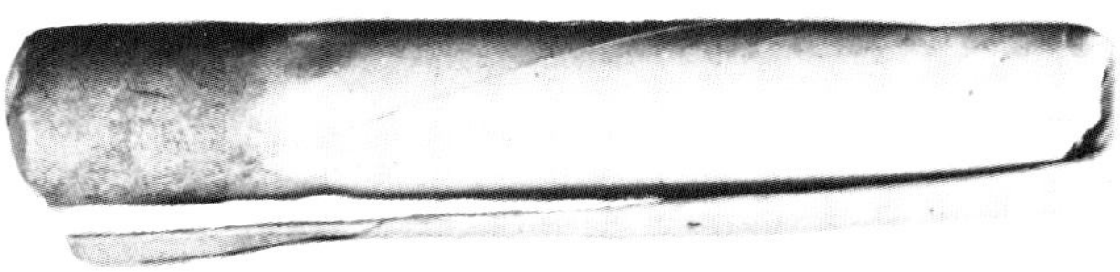

Cream-colored paper tube cemented over bullet base. Bullet type not discernible. Powder column closed by double folding tail which is nearly as long as entire cartridge.

Comments

These cartridges of the late muzzle-loading rifle period reflect a reduction in bore diameter. Aimed fire from accurate rifles ended the tactical use of massed bodies of infantry. With the range gradually being extended between the front lines of opposing ground forces, rifles firing the heavy, slow-moving projectiles were ineffective. Smaller-diameter projectiles with greater length propelled by proportionately larger powder charges evolved to cope with infantry units operating in skirmish lines at distances beyond the range of the earlier rifles.

These three specimens were among those in the White & Munhall ammunition collection and the identifications are those from the accompanying records. Verification could not be made from other sources. PR 13, however, does contain the deep-grooved compression bullet known to have been used by the Austrians.

PR 16 – 10.35mm (.407in) Swiss Federal Rifle, Model 1851

Description

White or buff paper tube cemented closed on base with double-folded tail, containing only a powder charge. A two-groove bullet moulded by the soldier was covered from base to center with a greased cotton patch tied on with fine string. (Drawing by Richard A. King)

PR 17 – 10.35mm (.407in) Swiss Federal Rifle, Model 1856

White or buff paper tube cemented over a lubricated two-groove pointed bullet. Powder charge separated from the bullet by an inner paper cylinder.

PR 18 – 10.35mm (.407in) Swiss Federal Rifle, Model 1863

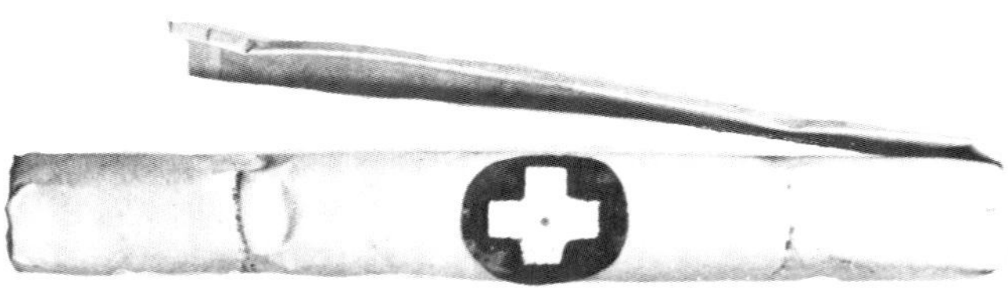

Essentially the same construction as above but slightly longer overall, and loaded with the Buholzer bullet which had two wide, shallow grooves and a base cavity, modified from the Minie type. The Swiss cross is in a red shield.

Comments

When the Swiss adopted the Federal rifle in 1851, they selected the smallest bore of any country in Europe. The bullet, however, was of considerably greater length for its weight than any other muzzle-loading bullet then in use. The Swiss rifle was known for its remarkable accuracy and range.

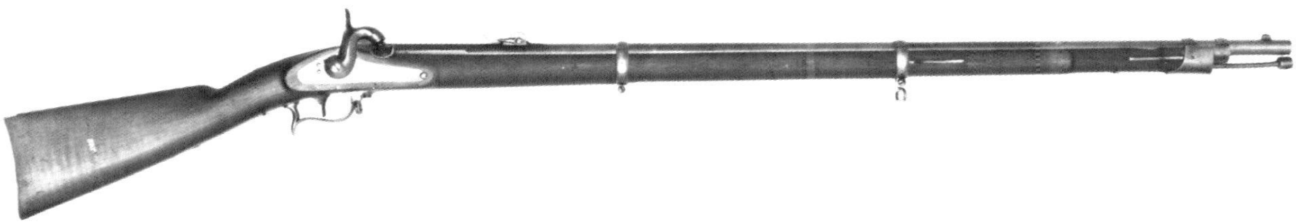

In 1847 the Swiss began extensive experiments with the various muzzle-loading rifles then in use in Europe and America. After trials lasting into 1850, the commission responsible for the work concluded that a relatively light-weight arm firing an elongated projectile 10.35mm (.407in) best suited the Swiss Army requirements. This rifle, called the "New Federal Stutzer" (provincial term for a rifled barrel gun), was adopted in 1850.

The remarkable feature of the Federal Rifle was its revolutionary small caliber, evidently patterned to some extent after the small-caliber long rifles used in America. The Swiss abandoned both the pillar breech and chambered rifle ideas developed in France, and neatly solved the problem of getting the bullet down on the charge without crushing the powder by making the ramrod with a swell at a point near its uper end which would butt against the muzzle and allow the rod to go only far enough into the barrel to seat the bullet just short of the charge.

The bore of the Federal Rifle was later enlarged slightly to 10.49mm (.413in) and the bullet increased in length. Ultimately, the pattern adopted had very shallow grooves and a modified hollow base similar to the Minie bullet which allowed the powder explosion to expand it into the rifling. Propelled by a 62 grain (4 grams) powder charge, this bullet gave greater range, accuracy, and penetration, and a flatter trajectory, than any other muzzle-loading rifle in use by any nation.

The alterations in bore size described above evidently are the reason why these Swiss rifles are seen identified as being from 10.3mm to 10.5mm. (Milwaukee Public Museum photo)

PR 19 – 18mm (.708in) Swiss Prelaz-Burnand Rifle, Model 1859 (altered from smoothbore musket)

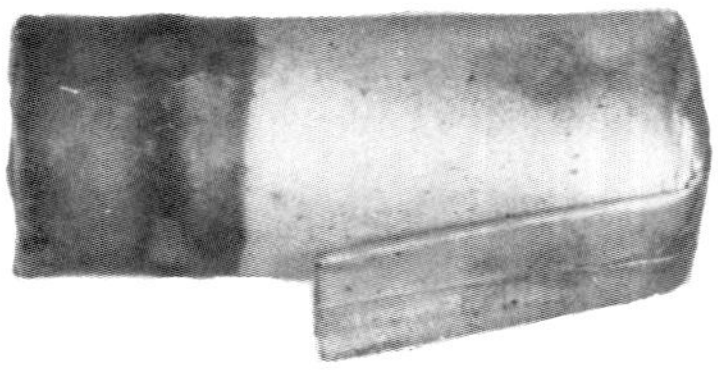

Description

White or buff paper tube folded into the base cavity of a single-groove, lubricated Minie-type bullet wrapped separately in paper. Powder column closed and tail folded as in the Federal rifle cartridge.

PR 20 – 18mm (.708in) French "Doremus" whole inserted Rifle Cartridge

Single-groove semi-pointed bullet wrapped in a thick paper patch with a charge of compressed black powder cemented to and filling the hollow base of the projectile. An additional strip of white paper covers joint of bullet and powder. The powder is covered with an outer paper cap which was removed before loading.

Comments

The Swiss Model 1842 rifle-musket employed the upper cartridge shown. The lower specimen is similar to such a cartridge shown on a cartridge board lithograph made up in the late 1800's by the French ammunition manufacturers Societe Francais des Munitions. The specimen is labeled "Systeme DOREMUS–cartouche & poudre comprimee" (referring to compressed powder). However, the specimen depicted on the lithograph lacks the protective cap on the powder. It remains to be estalished whether or not France used such ammunition as a regular issue item. It may have been only experimental.

PR 21 – 17.6mm (.692in) Austrian Rifle-Musket, Model 1847

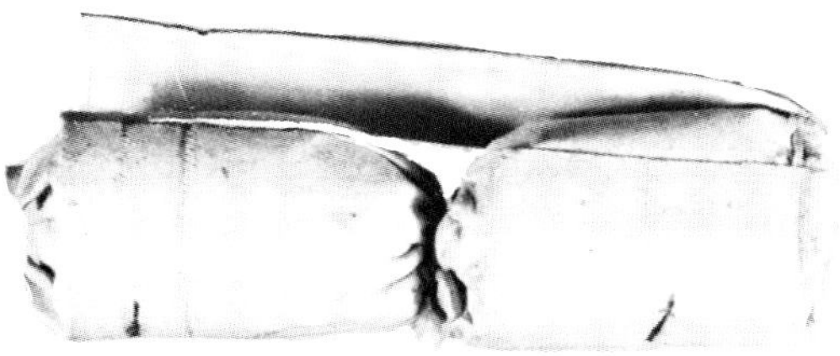

Description

White paper tube tied at the waist with string, folded around the base of a conical, grooved bullet without base cavity. Powder column closed by flattening tail, double folding, and laying it along case.

PR 22 – 16.8mm (.661in) Danish Restell, Model 1858

White or buff tube of used ledger paper employed to contain solid, conical bullet and powder charge separated by string tie. Case folded over bullet with lead showing, and powder column closed by gathering and tieing it around the neck of a paper mache "spoon" with hole in the center holding a percussion cap.

Comments

This cartridge-making technique had the advantage of strengthening the case body. Ammunition was subject to rough handling and tieing the bullet and charge apart not only separated them but allowed the case to bend without breaking apart. An Austrian reference indicates that the soldier pushed the bullet out of its paper covering to load. Both cartridges evidently were used in "tige" rifles in which the solid base of the bullet rested against a pillar extending up into the chamber. Several raps of the ramrod expanded the soft lead into the rifling.

The Danish specimen's curious spoon facilitated opening the powder column for loading, after which the soldier could direct the cap onto the nipple with this device. Pulling it away at right angles after capping left the primer on the nipple and the spoon was discarded.

PR 23(a) – .45in (11.43mm) English Whitworth, Hexagonal Bullet

PR 23(b) – .45in (11.43mm) English Whitworth, Hexagonal Bullet and Tube

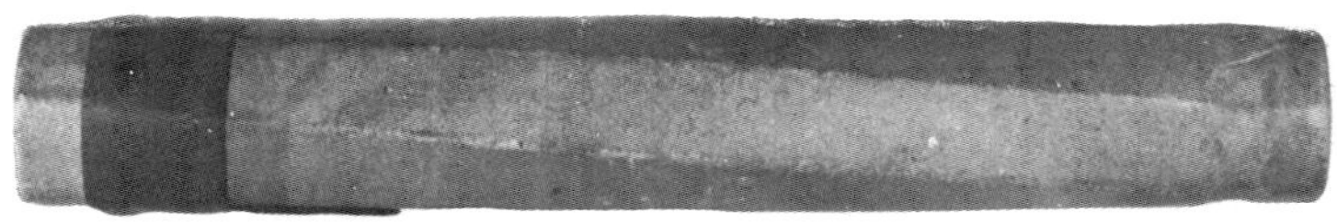

PR 23(c) – .45 Whitworth Charger and Bullets

Description

(a) White paper tube tied over the nose of a six-sided lubricated bullet. The paper is slit lengthwise in three places over the bullet, presumably to aid in its removal from the paper for loading. The powder column is closed with an additional wrap of paper and cemented around the end of the tube, the tail of which is twisted and pressed into the powder column.

(b) White or light brown cardboard tube formed with a hexagonal twist, containing powder charge with hexagonal bullet on top of the charge. This type was used with rifles counter bored at the muzzle to accept the base of the tube so the bullet would line up properly with the six-sided rifling. The bullet and powder charge were rammed into the barrel in one motion after the base tear tab was torn off.

(c) Not a cartridge, the inscription on this hollow brass Whitworth charger threaded at its closed end reads – "To screw on ramrod and convey a charge up the inverted barrel if desired." This device permitted loading powder without it touching the bore. The bullet used for such loading was carried in the cardboard container shown at the left of the bullets, together with a wax wad. The specimen on the right is the bullet contained in the specimens on page 46.

In 1854, Commander-in-Chief of the British Army, Lord Hardinge, asked Sir Joseph Whitworth, an engineer of some reputation, to investigate the subject of rifling. Sir Joseph was not encumbered with any pre-conceived notions on the matter since he had never worked with small arms.

He ultimately perfected a rifle with six-sided, or hexagonal, bore with spiral turn of one to 20 inches which fired a bullet moulded to mechanically fit this bore. In addition, he reduced the bore size to .45 inch (11.43mm) but retained the same 530 grain (34.3gm) bullet weight used in the Enfield rifle. This resulted in a much longer projectile with greater ballistic efficiency than the Enfield.

Sir Joseph's rifle markedly reduced the angle of elevation required for the .577in (14.65mm) Enfield rifle; i.e., the trajectory was considerably lowered. Moreover, the six-sided bullet precisely fitting the hexagonal bore vastly improved accuracy. Tests conducted in 1861 showed that the Whitworth would shoot into about a seven-inch circle at 300 yards. The Enfield could do no better than a 25-inch circle.

In 1863, the British Government contracted with Whitworth to supply his rifles and in 1864 he delivered some 8000. They were issued experimentally to certain regiments and remained in service until the Snider centerfire breech-loader adopted in September 1866 had replaced all muzzle-loaders. However, none of the Whitworth rifles saw action in the British Service. The Confederate forces purchased a number of them during the American Civil War and used them effectively as sharpshooter's rifles.

Powder fouling was a problem with the Whitworth. These deposits after a few shots tended to fill up the corners of the hexagonal rifling, making loading difficult. A scraper designed to fit the bore was a drastic and not very satisfactory answer.

Reference: "The Book of the Rifle," 1901, Hon. T.F. Fremantle, V.D., Major 1st Bucks V.R.C. (later Lord Cottesloe)
"19th Century Firearms," Maj. F. Myatt, MC, 1979

PR 24(a) – .45in (11.43mm) English Whitworth, cylindrical bullet

PR 24(b) – .45 (11.43mm) English Whitworth, cylindrical bullet and tube

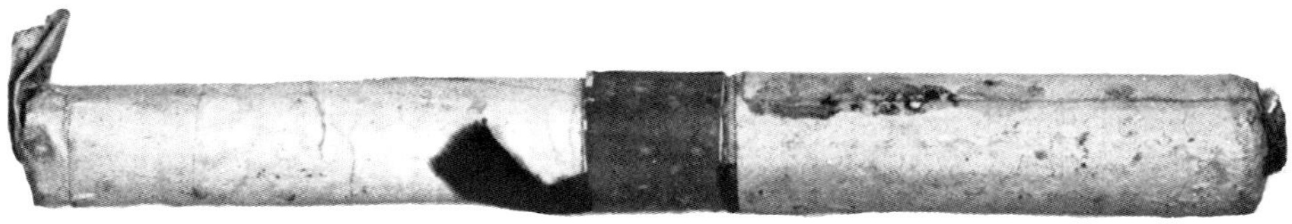

PR 24(c) – .45in (11.43mm) English Whitworth, cylindrical bullet, tube loader

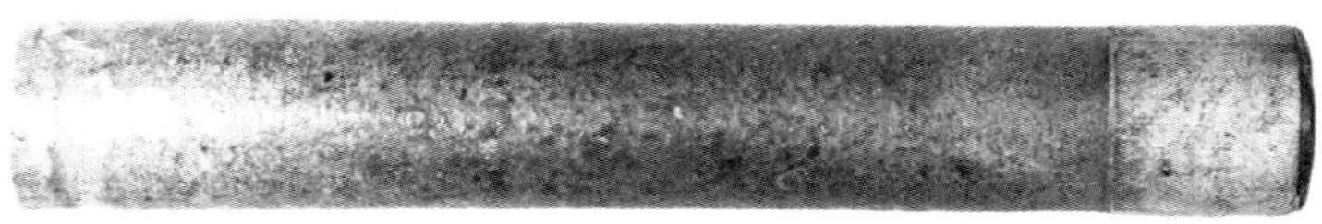

Description

(a) White cardboard tube containing a lubricated, cylindrical bullet and powder charge. A separate outer wrapping of paper is tied closed over the base of the bullet and the other end twisted closed over the powder column.

(b) A composite cartridge made of a cardboard tube containing the cylindrical bullet and a separate paper cylinder filled with powder inserted into the mouth of the bullet container. The two are joined in the middle by a cemented on strip of blue-green paper under which is affixed a fabric tear tab to facilitate opening and separating the two components. The cartridge is tied closed over the bullet and the tail is twisted to close the powder column. X-rays of this cartridge show a length of wooden stick within the powder charge.

(c) Brown cardboard cylinder containing powder and ball, constructed essentially the same as PR 23(b). The cartridge was used like a loading tube. The muzzle of the rifle barrel was enlarged for a short distance to permit this tube to be inserted. Then the powder and bullet could be rammed through the tube into the barrel in one motion.

PR 25 – 13.7mm (.539in) Austrian Von Lenk Guncotton Cartridge, 1861

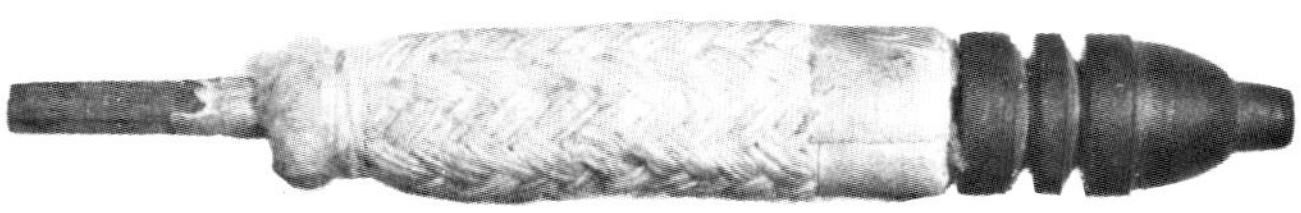

Description

White guncotton woven onto a stick which is inserted into a hole in the base of the two-groove compression bullet with projection on its nose intended to hold a cardboard washer larger than the bullet.

PR 26 – .69in (17.53mm) U.S. Mefford Sub-Caliber Projectiles, ca 1861-1865

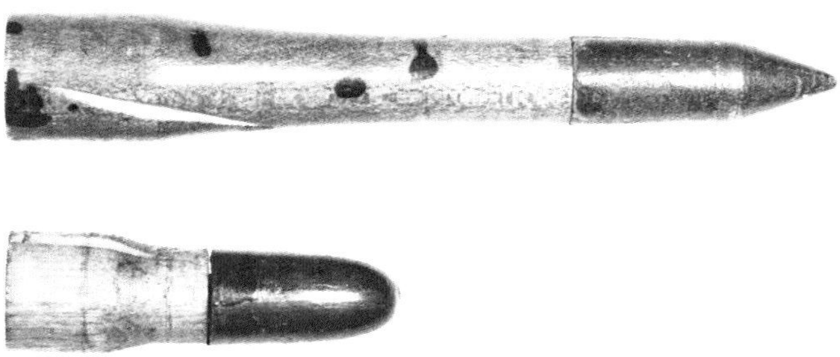

Lead conical or round-nose bullet attached to a wooden extension. The upper specimen, being just under bore size, presumably was loaded naked. The second is smaller in diameter and presumably was loaded with paper cartridge covering. These may have been fired in smoothbore arms as well as the rifle-musket.

Comments

Guncotton, a smokeless propellant invented by the Swiss Professor Schönbein, was applied by Austrian army officer Baron Van Lenk, to small arms ammunition. He wove the guncotton in the specimen shown onto the stick in order to control the burning rate. The cartridge, whose bullet was less than bore size, was dropped down the elevated rifle muzzle. The soldier rapped the butt of the arm on the ground two or three times to seat the base of the stick in a depression made for it in the bottom of the chamber. This secured the cartridge in the chamber. The cardboard washer fitted over the nose of the bullet was stripped off at the muzzle and the soldier pressed it into a recess machined into the muzzle. This served as a reminder that the rifle was loaded and prevented accidental loading of two cartridges. The washer, of course, was blown off when the rifle was fired. U.S. forces experimented with this cartridge in the .58 rifle but guncotton ultimately was determined unsuitable for small arms because of its violent explosive characteristics.

The Mefford sub-caliber projectiles were reportedly devised during the American Civil War to obtain higher velocity for penetrating light fortifications. However, Colonel Lewis illustrates a specimen of the Mefford cartridge loaded with buck and ball.

PR 27 – .58in (14.73mm) U.S. Rifle, ca. 1861-1865

Description

Black powder compacted in dry condition within a mold by pressure to condense and form it to the desired shape. The charge is cemented to a conical, hollow-base, two-groove bullet. Collodion (nitrocellulose dissolved in ether and alcohol forms the varnish to waterproof the charge and facilitate ignition.

PR 28 – .69in (17.53mm) U.S. Rifle-Musket, ca. 1861-1865

Same construction as that shown above.

Comments

These whole-inserted muzzle-loading cartridges, similar to that shown at PR 20, were made under the Doremus and Budd patents. Hazard Powder Company manufactured this type of ammunition during the Civil War for Colt, Sharps, and Hall breech-loading arms as well as for the muzzle-loaders. They obviated the necessity for separate loading of powder and bullet but, because of their fragility, they had to be carefully wrapped individually. Thus, the soldier still had to remove the cartridge from its wrapping and container.

PR 29 – .54in (13.72mm) for U.S. 1841 Rifle, ca. 1861-1865

Description

Molded black powder charge with a wrapping of silk fabric cemented around the powder and the lower portion of the projectile.

PR 30 – .58in (14.73mm) for U.S. Rifle, Model 1855 and others, ca. 1861-1865

Same as above but the wrapping of silk fabric not as visible.

PR 31 – .69in (17.53mm) U.S. Rifle-Musket,ca. 1861-1865

Another constructed on the same system as the above, but silk fabric wrap not as visible in this photo.

Comments

The Bartholow patent covering these cartridges called for the mixture of finely divided ingredients of black powder with tincture of shellac which was then molded into a solid explosive casting. The solid charge was then cemented to the base of the bullet with marine glue or other adhesive substance. The silk fabric, not claimed in the patent, was intended to give further rigidity to the juncture of charge and projectile. The silk was consumed with the powder. PR 31 also is found with buck and ball load.

PR 32 – 14.73mm (.58in) Unknown,
ca. 1850(?)

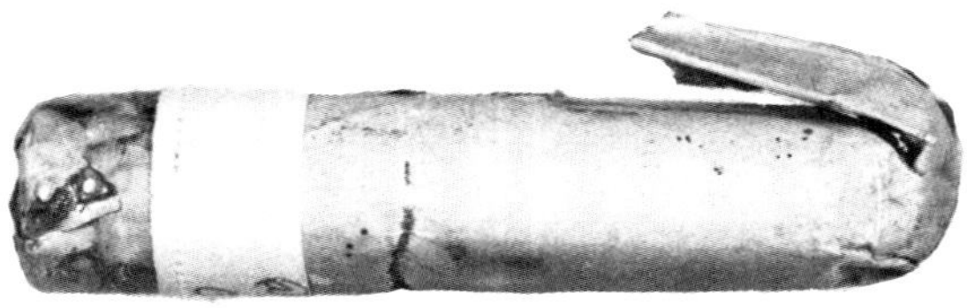

Description

White paper case cemented closed over a hollow-base expansion bullet with iron cup (bullet takes a magnet.) Tail folded and laid along the case. Written on the side of the case in French is "avec culot" (with cup).

Comments

The specimen above is representative of the early French experiments with the metal expanding cup in the hollow base of the bullet designed to spread the projectile into the rifling from the force of the exploding powder charge. The cup, or "culot," was subsequently found to be unnecessary and was abandoned. The rifle for which this cartridge was made could not be determined. Since other countries also experimented with the expanding cup, it may not be French.

PERCUSSION MUZZLE-LOADING RIFLE AND CARTRIDGE DATA

No.	Bore	Model and Date	Bullet Diameter	Bullet Weight	Powder Charge	Cartridge Length	Sources
PR 1	.54in 13.72mm	U.S. Harpers Ferry, Model 1841	.551in 14mm	400grs 25.92gm	50-60grs 3.2-3.9gm	2.35in 59.69mm	Specimen—Lewis
PR 2	.69in 17.53mm	U.S. Rifle-Musket, Model 1842 (altered)	.632in 16.05mm	700grs 45.36gm	110grs 7.13gm	3.20in 81,28mm	Specimen—Lewis
PR 3	.58in 14.73mm	U.S. Rifle-Musket, and Models 1841 and 1855	.591in 15.01mm	555grs 35.96gm	60grs 3.9gm	2.40in 60.96mm	Specimen—Lewis
PR 4	.54in 13.72mm	Confederate States of America, ca 1861-1865	.554in 14.07mm	400grs 25.92gm	50-60grs 3.2-3.9gm	2.6in 66.94mm	Specimen
PR 5	.69in 17.53mm	CSA Rifle-Musket	.670in* 17.02mm			2.22in 56.39mm	Specimen
PR 6	18mm .708in	French, Model 1842	17.65mm .695in			69.85mm 2.75in	Specimen
PR 7	.69in 17.53mm	Mexican Rifle-Musket, ca 1850-1860	.698in 17.73mm			2.45in 62.23mm	Specimen
PR 8	17.5mm .689in	Belgian Minie Rifle, ca 1863	17.22mm .678in			66.29mm 2.61in	Specimen
PR 9	.702in 17.83mm	British Minie Rifle, Model 1851	.700in 17.78mm		68grs 4.41gm	2.45in 62.23mm	Specimen
PR 10	.577in 14.65mm	British Enfield Rifle, Model 1853	.568in 14.43mm	530grs 34.34gm	68grs 4.41gm	3.12in 79.25mm	Specimen—Lewis
PR 10(a)	.577in 14.65mm	British Enfield Rifle, Model 1853	.568in 14.43mm	520grs 33.70gm	61.5grs 3.99gm	2.90in 73.66mm	Specimen—Schon
PR 10(b)	.577in 14.65mm	British Enfield, Model 1853	.568in 14.43mm	530grs 34.34gm	68grs 4.41gm	3.12in 79.25mm	Specimen—Lewis

PERCUSSION MUZZLE-LOADING RIFLE AND CARTRIDGE DATA

No.	Bore	Model and Date	Bullet Diameter	Bullet Weight	Powder Charge	Cartridge Length	Sources
PR 11	14.65mm .577in	Turkish Enfield, Model 1860	14.61mm .575in			66.80mm 2.63in	Specimen
PR 12	15.24mm .600in	Russian "6-Line" Rifle, Model 1857	15.24mm .600in			60.96mm 2.40in	Specimen
PR 13	13.9mm .547in	Austrian Rifle-Musket, Model 1855	13.97mm .550in	26-28gm 401-432grs	4-4.5gm 62-69grs	66.8mm 2.60in	Specimen—Elgger
PR 14	13.9mm .547in	Prussian Lorenz Rifle, Model 1864	13.74mm .541in	26-28gm 401-432grs		66.04mm 2.60in	Specimen—Elgger
PR 15	12.5mm .492in	Dutch Rifle, Model 1863	13.54mm .533in			79.50mm 3.13in	Specimen
PR 16	10.35mm .407in	Swiss Federal Rifle, Model 1851	10.20mm* .402in	16-17gm 247-262grs	3.8-4.2gm 58-65grs	63.50mm 2.50in	Schneider and am Rhyn
PR 17	10.35mm .407in	Swiss Federal Rifle, Model 1856	10.2mm* .402in	16-17gm 247-262grs	4gm 61.7grs	82mm 3.2in	Specimen
PR 18	10.49mm .413in	Swiss Federal Rifle, Model 1863	10.2mm* .402in	18.6gm 287grs	4gm 61.7grs	86mm 3.38in	Specimen—Mordecai after J. Schön
PR 19	18mm .708in	Swiss Prelaz-Burnand Rifle, Model 1859	17.8mm .718in	36gm 556grs	4.5gm 69grs	49.5mm 1.95in	Specimen
PR 20	18mm .708in	French "Doremus," ca 1859	18.24mm .718in	36-38gm 556-586grs		43.18mm 1.70in	Specimen

PR 21	17.6mm .692in	Austrian Rifle-Musket, Model 1847	18.24mm .725in	40gm 617grs	4.5gm 69grs	60.45mm 2.38in	Specimen—Elgger
PR 22	16.8mm .661in	Danish Restell, Model 1858	16.64mm .665in			90.68mm 3.57in	Specimen
PR 23(a)	.45in 11.43mm	Whitworth Hexagonal, Model 1854	.442in* 11.23mm	530grs 34.34gm	85grs 5.51gm	4.57in 116.1mm	Specimen—Lewis
PR 23(b)	.45in 11.43mm	Whitworth Hexagonal Tube, Model 1862	.442in* 11.23mm	530grs 34.34gm	85grs 5.51gm	3.88in 98.55mm	Specimen—Lewis
PR 24(a)	.45in 11.43mm	Whitworth Cylindrical, ca 1862	.442in* 11.23mm	530grs 34.34gm	85grs 5.51gm	4.8in 121.9mm	Specimen
PR 24(b)	.45in 11.43mm	Whitworth Cylindrical, ca 1865	.442in* 11.23mm	480grs 31.1gm	75grs 4.86gm	4.8in 121.9mm	Specimen—Lewis
PR 24(c)	.45in 11.43mm	Whitworth Cylindrical Tube, 1862	.442in* 11.23mm	530grs 34.34gm	85grs 5.51gm	3.63in 92.2mm	Specimen—Lewis
PR 25	13.7mm .539in	Austrian Von Lenk Guncotton, 1861	13.41mm* .528in	31gm 478grs		90.68mm 3.57in	Specimen
PR 26	.69in 17.53mm	U.S. Mefford sub-cal., ca 1861-1865	.674-.462in** 17.1-11.7mm			4.19in 106.4mm	Specimen
PR 26(a)	.64in 16.26mm	U.S. Mefford sub-cal., ca 1861-1865	.633-.513in 16-13mm	360grs 23.33gm		1.95in 49.53mm	Specimen

PERCUSSION MUZZLE-LOADING RIFLE AND CARTRIDGE DATA

No.	Bore	Model and Date	Bullet Diameter	Bullet Weight	Powder Charge	Cartridge Length	Sources
PR 27	.58in 14.73mm	U.S. Rifles, ca 1861-1865	.577in* 14.65mm	490grs 31.75gm	65grs 4.21gm	1.82in 46.23mm	Specimen—Lewis
PR 28	.69in 17.53mm	U.S. Rifle-Musket, Model 1842	.685in* 17.40mm	745grs 48.6gm	70grs 4.54gm	1.75in 44.45mm	Specimen—Lewis
PR 29	.54in 13.72mm	U.S. Rifle, Model 1841	.549in 13.94mm	450grs 29.2gm	50grs 3.24gm	1.69in 49.93mm	Specimen—Lewis
PR 30	.58in 14.73mm	U.S. Rifles, ca 1861-1865	.576in* 14.63mm	490grs 31.75gm	65grs 4.21gm	1.77in 49.96mm	Specimen—Lewis
PR 31	.69in 17.53mm	U.S. Rifle-Musket, Model 1842	.685in* 17.40mm	750grs 48.6gm	70grs 4.54gm	2.14in 54.36mm	Specimen—Lewis
PR 32	14.83mm .58in	French Rifle (?), ca 1851	15.19mm .598in			72mm 2.83in	Specimen

Notes: *Actual bullet diameter. Otherwise, bullet diameter taken over the paper case. The actual diameter of the projectile in these specimens will be from .10 to .20 thousandths of an inch less, depending on paper thickness.

**The first dimension for the two Mefford sub-caliber projectiles is the wood diameter; the second, the diameter of the lead point.

7 THE PERCUSSION BREECH-LOADER

The percussion breech-loader (also capping breech-loader) is a firearm charged at the breech and fired by a percussion cap, tape primer, or disc primer which is positioned on an external nipple. The explosion of the fulminate in the cap when struck by the hammer travels down the ignition vent leading from the hollow nipple to reach the powder charge.

The earliest of these guns, the American Jenks, was loaded with a round ball and loose powder through an opening in the top of the breech and barrel assembly when a lever was drawn up and back. Thus, no cartridge was made for it. Others used paper cartridges which were opened and the powder poured in the chamber followed by the bullet. Later types used paper or combustible cartridges loaded whole. In these, the percussion flame had to penetrate the case material to ignite the powder and the cartridge case was burned with the exploding powder. The final and most sophisticated actions employed metallic cartridge cases perforated at the rear to admit the percussion flame to the powder charge.

The tip-up chamber loading system of the Austrian Gorla/Crespi and Hall flintlock arms emerged in the Norwegian Government's underhammer breech-loader and the Hall was successfully converted to percussion. Edward Lindner applied the principle to his 1859 action. Between 1840 and 1862, dozens of other actions were developed in America and Europe. They included top-lever opening, tilting barrel, rotating breechblock, sliding bolt, sliding barrel, tilting breechblock, dropping breechblock, and others.

Cartridge Case Types

Percussion breech-loading ammunition shows intense experimentation with case material and construction. In the following pages, the reader will find cartridge cases constructed of paper, linen, silk mesh, animal intestine or other membrane, hard rubber, brass foil, and drawn brass.

It should be noted that the American Hall flintlock rifle converted to percussion and accepted by the U.S. Army in 1834 was the first percussion arm generally adopted by a military power.

PBL 1 – .64in (16.25mm) U.S. Hall Carbine of Musket Caliber, 1839

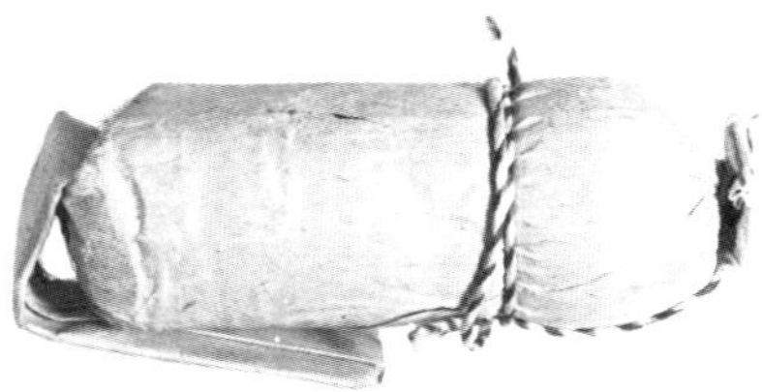

Description

White paper tube tied over and below a round ball with red and white string. Powder column closed with a double fold of the tail to the center and laid along the case.

PBL 2 – .52in (13.21mm) U.S. Hall Rifle and Carbine, 1834-1842

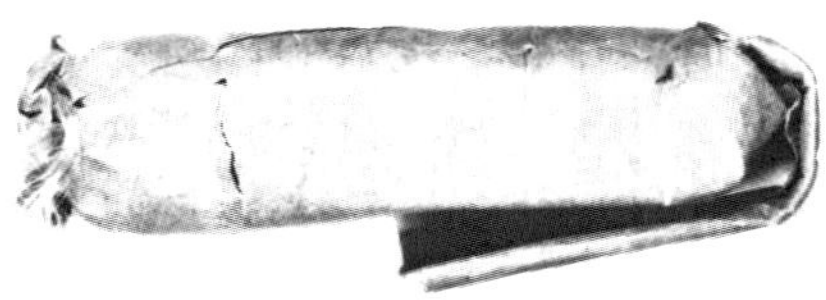

Brown paper tube cartridge constructed as above but using tan string for the ligature.

PBL 3 – 16.89mm (.665in) Norwegian Underhammer Rifle, Model 1842

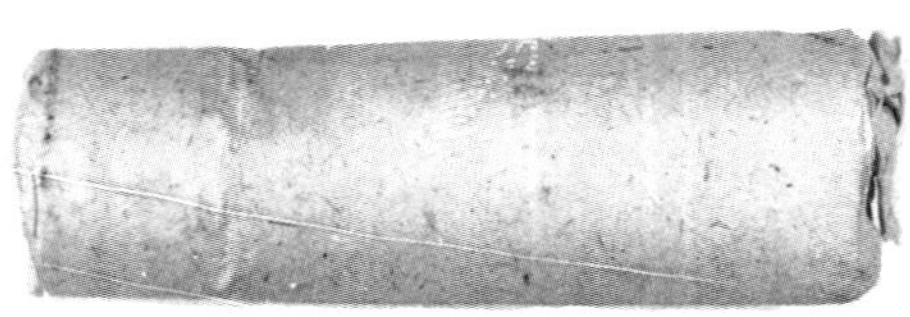

White paper folded closed over base of a conical bullet. Powder column twisted closed and tail pressed into base cavity.

Comments

The Hall breech-loading carbine cartridges represent the simplest form of paper-case ammunition for such arms, differing from the muzzle-loading type only by the smaller powder charges. Although the flintlock Hall had a rifled bore, both it and the later percussion models used only round ball ammunition. The ball in these two cartridges was slightly larger than the bore, so there was no "windage" between ball and bore. However, a few smoothbore Hall carbines were made during the percussion era where ball to bore fit was not so complete.

The Norwegian underhammer rifle cartridge followed the earlier-developed principle of conical ball below the powder charge. The mouth of the powder column was torn open, the powder poured in the up-tilted chamber, and the bullet inserted on top of the charge. A later version of this cartridge was made with the reverse-taper paper powder container tied to the conical ball and the entire cartridge, paper and all, was loaded in the chamber.

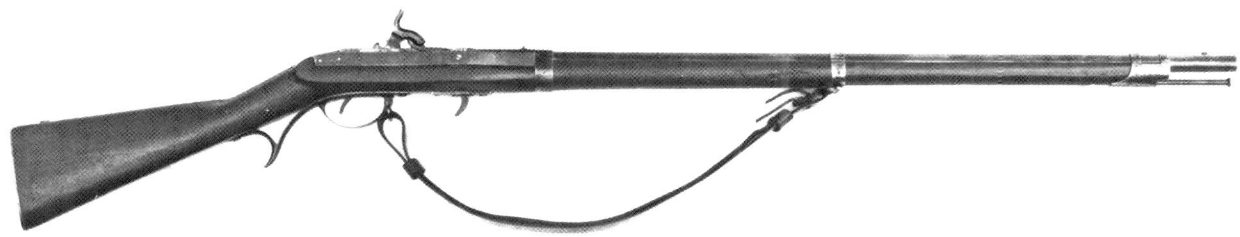

A .52 caliber Hall rifle converted from flintlock to percussion. The original hammer was replaced with the percussion type, the pan and pan cover assembly removed, and the original vent plugged. The top of the breechlock was ground down to accept a plate with the rear edge tipped up. Finally, the plate and breechlock were bored through into the chamber and threaded. A percussion nipple was screwed in just ahead of the top rear edge of the added plate which served as a shield to deflect flying bits of percussion cap from the shooter's face. This rifle is operated by an underlever visible only when the chamber is snapped down in line with the barrel. This original operating lever was found objectionable because it could accidently open the action on the march. If the soldier did not notice this, he could fire the rifle with the breech open. A later type employed a side lever patented by Savage & North. To load the Hall, the breechlock was lifted, loose powder poured in from the paper cartridge, and the ball pressed in on top of the charge. The breechlock was closed down, a percussion cap placed on the nipple, and the rifle was ready to be fired. (Steve Brenger Collection)

The 1842 underhammer rifle of Norway was a chamber loader like the Hall. However, the percussion nipple was on the bottom of the chamber and fired by the hammer situated ahead of the trigger guard. The curved piece ahead of the hammer is the spring which drove the hammer onto the nipple when the trigger was pulled. Several models were made in 16.89 (nominally 17mm) and 12mm and they were later converted to use rimfire ammunition (See Chapter 11). (Drawing from Rodolphe Schmidt's "Les Armes A Feue Portatives," 1877)

PBL 4 – .56in (14.22mm) U.S. Merrill Carbine, Model 1859

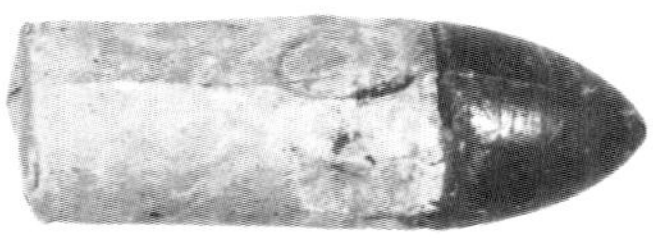

Description

Tan paper case cemented around base of a conical, grooved bullet. Powder column closed by folding and cementing base.

PBL 5 – .63in (16mm) Unknown

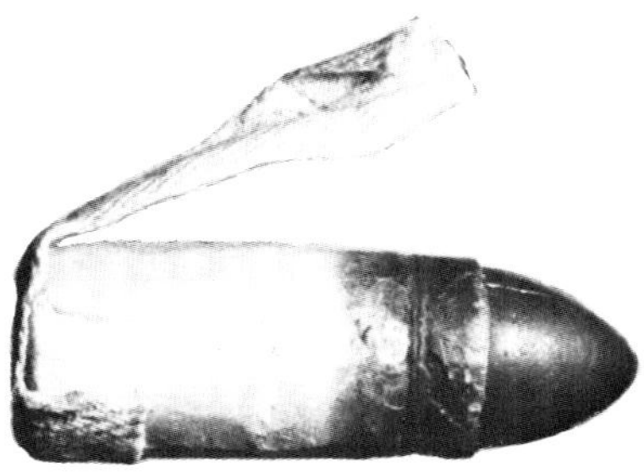

White paper case tied to a conical bullet. Powder column closed by folded and cemented base with central perforation. Linen tail cemented over base.

Comments

The Merrill cartridge was inserted whole in the chamber and the flame from the exploding percussion cap penetrated the paper case to ignite the powder charge. The case was expected to be consumed by the burning powder charge.

The unidentified specimen, made for a percussion breech-loader with ignition orifice leading to the base of the case, was also self-consuming. The linen tail was stripped off the base to expose the perforation before loading.

Neither of these cartridges contributed anything to sealing, or obturating, the joint between breech and barrel.

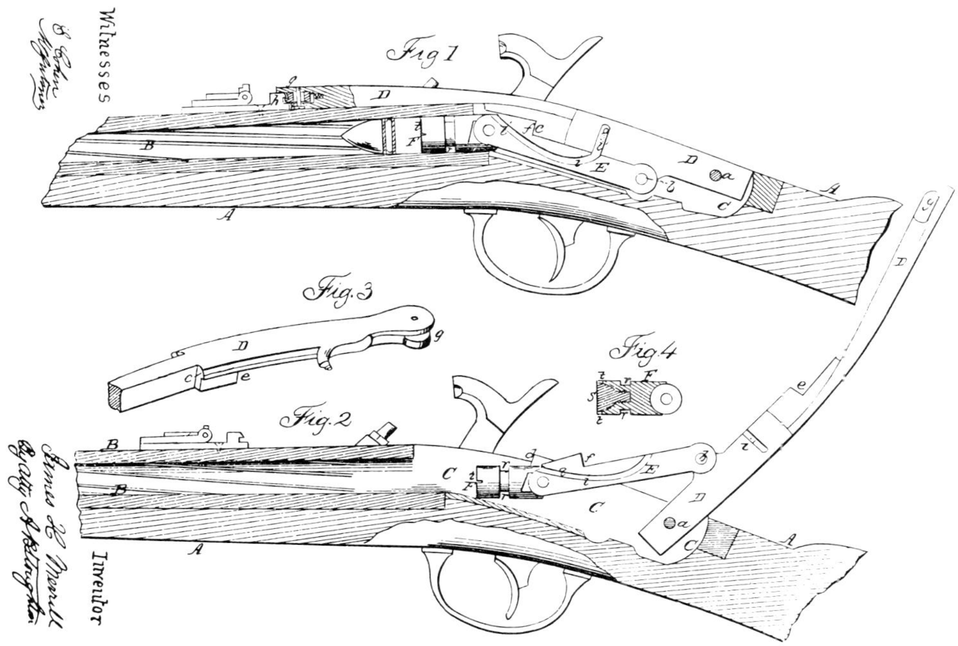

The Union purchased 14,495 carbines during the American Civil War made with this breech-loading action patented by J. H. Merrill in April 1861. To open the action, the operating lever was released from its notch to the rear of the back sight and the lever was drawn up and back. This withdrew the bolt from the chamber. The paper-case cartridge was inserted in the chamber and the lever pushed down to lock position. This moved the piston bolt into the chamber, driving the cartridge in with the bullet seated against the rifling. The face of the bolt was hollowed out to form a cavity which was filled with copper. The copper was expected to expand under the force of the exploding powder charge to expand the bolt and form a gas seal.

An earlier Merrill model patented in 1856 had a rotating chamber with operating lever lying along the tang when closed. It was lifted forward to open for loading. Both of these arms were related in action principle to the Jenks breechloader, first made with flintlock ignition and conventional hammer, and then in percussion with a side, or "Mule Ear," hammer. Raising the operating lever exposed an opening into the chamber which was charged with naked lead ball and loose powder. The U.S. Government provisionally adopted the Jenks in 1839 for tests. (Illustration after an original patent drawing)

Description

PBL 6 – .56in (14.22mm) U.S. Sharps Rifle, 1858

Grey paper case cemented and string tied around heel of a multi-groove conical lead bullet. Powder column closed by flattening and double folding tail. Also found in .44 and .36 calibers for sporting rifle.

PBL 7 – .52in (13.22mm) U.S. Sharps Rifle, Model 1861

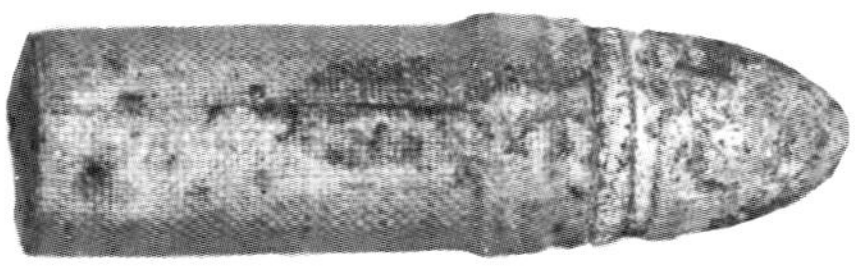

White linen tube cemented around base of a grooved conical bullet. Powder column closed with a circular patch of nitrated paper.

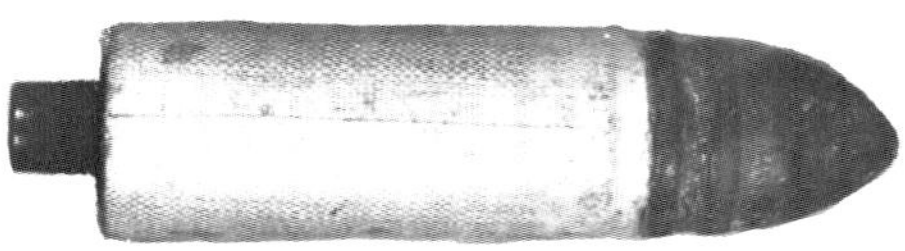

(b) Variant of above with the percussion cap cemented to the base of the case, presumably to be handy for the soldier to pick off with his fingers during loading.

PBL 8 – .52in (13.22mm) U.S. Sharps Carbine, Model 1861

White linen tube with shorter powder column than above and with dark patch cemented over and above the base. This is a commercially-manufactured cartridge for sporting use.

Comments

PBL 6 is representative of the first Sharps rifle and carbine made longer than the chamber. The sharp edge of the closing breechblock lifted to cut off the projecting tail, thus exposing the powder to the ignition orifice. Loose powder was often left on top of the action by this opening of the case. The powder gases which inevitably escaped to some degree between the joint of the breechlock and barrel often ignited this powder which could spew into the shooter's face.

This troublesome problem was solved by employing a durable linen case with thin nitrated paper covering the powder which was easily penetrated by the exploding percussion cap. Both types of cartridges were expected to consume the case material but embers of burning paper or linen could remain in the chamber after firing, creating some hazard if the arm was immediately reloaded. However, the residue was normally pushed out ahead of the cartridge being loaded.

Sharps New Model 1859 military rifle, caliber .52, made with the disc priming magazine visible at the top of the lock plate beneath and behind the hammer. Its serial number 56659 places it among the two thousand with double set triggers issued to the United States Sharpshooters commanded by Colonel Hiram Berdan during the American Civil War.

Sharps rifles are loosely divided into the early slant breech models and the straight breech models of which this is an example. The trigger guard operating lever drawn down and forward pulled the breechblock down in its mortise to expose the chamber. The paper or linen cartridge was pushed into the chamber and the trigger guard operating lever drawn back to close the action. Despite the use of platinum or copper gas seal rings on the face of the breech, this rifle was known for its tendency to leak gas at juncture of breechlock and barrel.

The disc primer mechanism was loaded with small flat pellets containing fulminate mixture where they were contained by the priming well cover and held under forward pressure by a magazine spring. Cocking the hammer and pulling the trigger to let it fall forced a pellet feeder forward, ejecting the top pellet from the magazine. This action and the fall of the hammer were timed so precisely that the hammer nose caught the pellet right on the nipple, crushing it and exploding the fulminate within. The pellet did not have far to travel from the magazine to nipple and the system evidently worked well, though some authors have stated that windy weather affected its reliability. This mechanism first appeared on the Model 1852 Sharps. In 1859, a patented priming magazine cutoff was added to allow the shooter to keep the disc primer magazine in reserve while he positioned percussion caps on the nipple by hand. The Maynard tape priming mechanism was also used on the early Sharps models.

The first 1849 Sharps rifle ushered in a long line of military, hunting, and target rifles, the last being made in the 1880's. Despite its inability to entirely seal in the exploding powder gases, the combustible cartridge Sharps was one of the best rifles of the Civil War and was widely copied, particularly in Europe. The vertical sliding block system invented by Christian Sharps is still used in the finest sporting single-shot rifles. (Rifle illustrated from the Steve Brenger collection)

Reference: "Sharps Firearms," Frank Sellers, 1978.

PBL 9 – .54in (13.72mm) U.S. Starr, 1858

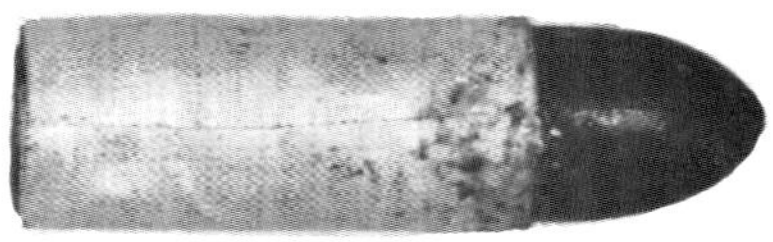

Description

White linen case loaded with a two-groove lead bullet, grooves within the case. Powder column closed with a disc of blue paper as made by St. Louis arsenal.

PBL 10 – .54in (13.72mm) U.S. Lindner for 1859 Alteration of Model 1842 Muzzle-loading Rifle

Tan paper tube cemented and tied to a grooved conical bullet. Powder column tapered to rear and closed with a stopper of yarn or roving cut in short sections. The yarn is secured in this specimen with a wrapping of thread around the case.

PBL 11 – .58in (14.73mm) U.S. Lindner Carbine, Model 1859

Same construction as above except the case mouth is cemented over all the bullet grooves. The powder column is closed with the yarn stopper held in place by a wrap of paper cemented around the base of the case. A variation exists with the case cemented on below both grooves in the bullet.

Comments

The Starr cartridge varies only in details from the Sharps linen specimens shown previously. Most were made with white paper closing the powder column.

Edward Lindner, a German-American living in New York, was issued a patent for his yarn-stopped cartridge in July 1861 to go with his 1859-patented tip-up chamber breech-loading rifle action. The patent describes the cartridge as follows: " . . .a ball cartridge in which the powder in the rear of the ball is contained in a paper cone twisted round a soft stopper that may be drawn out by the teeth when loading the gun. . .no powder is lost by careless tearing nor in pouring the powder into the barrel, because the end of the case is left even and of a smaller size than the bore of the gun. . ." From this it is evident that Lindner envisioned his cartridge as suitable for muzzle-loaders as well as his breech-loader. New York Shot & Lead Co. made this ammunition. (Research courtesy of Paul Klatt and Dean Thomas.)

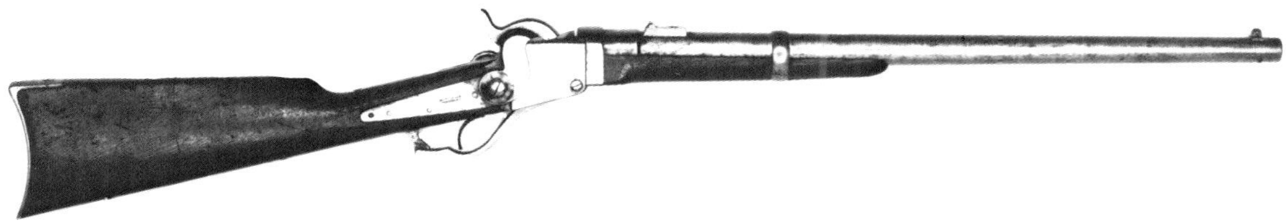

This percussion carbine patented by Eben T. Starr on September 14, 1858 and manufactured by him at Yonkers, N.Y. opened for loading when the trigger guard lever was depressed. The brace connected to the breechlock by means of a link drew it downward and pivoted it to the rear simultaneously. The face of the breechlock was countersunk annularly so as to give a conical bearing against the chamber. Over 25,000 of these arms were purchased by the Union.

The U.S. Government employed Lindner's 1859 action to convert a few .54 and .58 caliber percussion muzzle-loaders to breech-loaders. Amoskeag Mfg. Co., Manchester, New Hampshire, manufactured 892 carbines using the same action for the Union during the Civil War.

To operate this action, the handle attached to a locking sleeve covering the joint of breechblock and barrel was rotated up and to the left, or counterclockwise. The sleeve was cut away enough to allow the chamber to pass upward through the cut. The chamber, under pressure from a spring bearing against its underside, then lifted up at a 45 degree angle. After loading the hollow, uplifted chamber with the type cartridge shown at left, the soldier forced the chamber down manually in line with the barrel and rotated the sleeve by the handle back clockwise. The sleeve rode in a groove in the head of the breechblock cut at an angle so that, when turned closed, it cammed the breechblock against the rear face of the barrel to form a gas seal. (Milwaukee Public Museum photos)

PBL 12 – .58in (14.73mm) Unknown

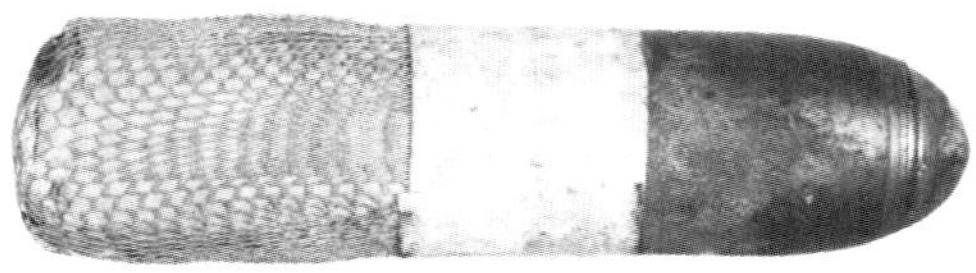

Description

Tan paper case covered with fine silk mesh cemented to a grooveless, conical bullet with a wrap of pink paper. Powder column closed by folding case paper and cementing, supported by the silk bag.

PBL 13 – 12mm (.472in) Manceaux & Viellard, ca. 1861

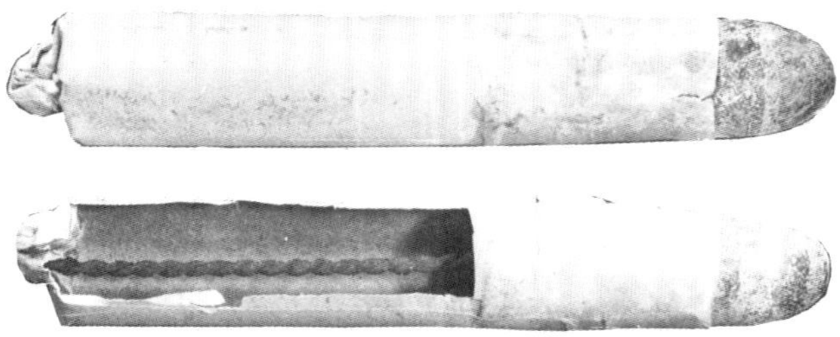

Paper case cemented around lower section of a grooveless, round-nose bullet. A hard fabric wad at the base of the bullet held one end of a twisted length of brass wire which passed through the powder charge. The other end was threaded through the base of the paper case and twisted to close the powder column.

Comments

The unknown specimen above evidently was an attempt to ensure ignition of the powder charge through the very thin, fragile paper and reinforcing silk mesh. The mesh is smaller than the powder grains which had already begun to spill from the broken paper case when the specimen was found. The mesh would not impede ignition and both being very thin would probably burn completely with the powder charge.

The French Manceaux & Viellard cartridge was a novel attempt to dispose of unburned paper case material. When the bullet and wad were propelled down the barrel by the exploding powder charge, the wire embedded in the wad was expected to carry with it the unburned paper remaining at the base of the case. The device suggests that unburned material left in the chamber of the rifle caused problems. The specimen on page 96 (PBL 43) appears to have included this same device in somewhat different form.

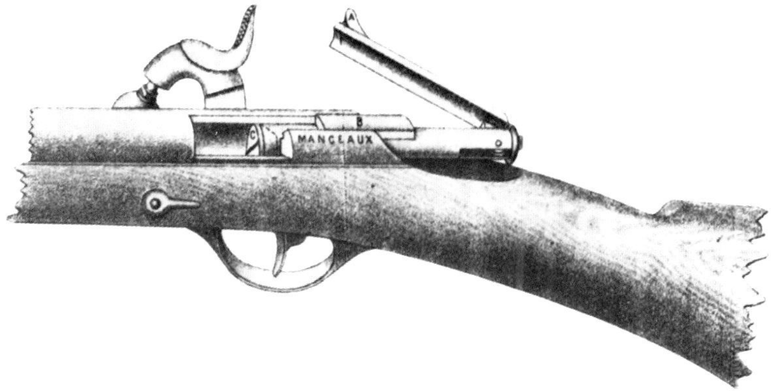

The Manceaux & Viellard rifle developed in France was constructed around the sliding door bolt action which had a hinged latch fastened to the rear of the bolt. To operate, this lever arm was raised from its closed position lying along the length of the bolt and the bolt rotated a quarter turn to disengage the locking surface (B) from the shoulder of the receiver housing. Then the bolt could be drawn back and a cartridge inserted in the chamber. Pushing the bolt home, turning it a quarter to the left, and laying the lever arm along the bolt completed the operation. The percussion cap fired through the ignition orifice into the center of the paper cartridge. A cone-shaped obturating plug, evidently of hard rubber or other elastic substance, was inset into the face of the bolt (C).

This rifle was experimental only. French military authorities did not adopt it. (Drawing from Rodolphe Schmidt's "Les Armes A Feu Portatives," 1877)

PBL 14 – .577in (14.65mm) English (Hayes Patent) for Mont-Storm Carbine, 1860

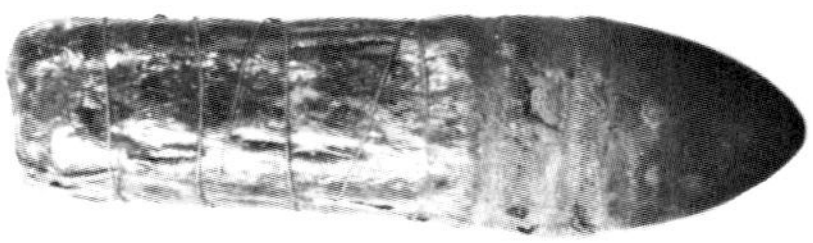

Description

Animal intestine sack filled with black powder and cemented around the lower circumference of a grooved, conical bullet. Powder container varnished and wrapped in opposite directions with thread.

PBL 15 – .577in (14.65mm) English Prince's Rifle, 1859

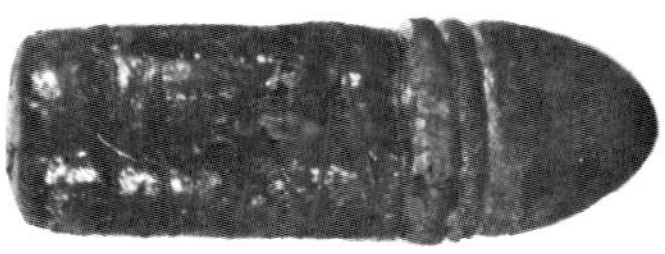

Same construction and materials as above, differing only in the shape of the bullet and length of the powder charge.

Comments

Captain John M. Hayes, Royal Navy, received a patent for this type cartridge in England on September 4, 1856. It specified as case material the skin or membrane prepared from the gut of animals such as pigs, birds, or reptiles instead of paper, with the strengthening netword of thread. To further protect the cartridge from damage prior to use, it was enclosed in a paper envelope with cemented seams and a fabric tear tab for opening. The envelope was covered by a different patent taken out by Samuel Colt and William Thomas Eley. The Hayes patent skin cartridges are far more common in pistol and revolver sizes than in these larger rifle rounds. Eley, Broux & Mall, and H. Gladstone & Co. of England manufactured such ammunition.

PBL 14 is commonly referred to as for the Enfield muzzle-loaders. However, it appears unlikely that this particular cartridge could be loaded in a .577-bore rifle since the diameter of the bullet slightly exceeds that (See dimensions table at end of chapter). Available references establish that this type ammunition was used in the Mont-Storm.

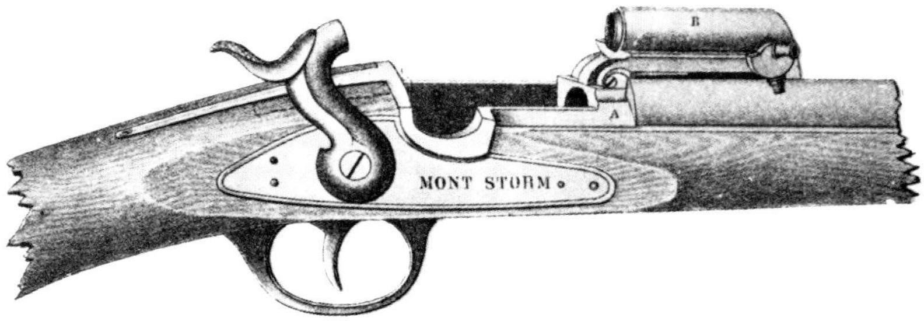

The Montgomery-Storm, or Mont-Storm as it generally is known, was a chamber loader with breechblock hinged at the front so that it opened by tipping up and over the barrel. The cartridge was inserted base first and the chamber returned to the closed position in line with the barrel. The ignition orifice was at the rear of the chamber. This rifle was not well liked because the shooter might insert a cartridge into the chamber bullet first. This put it out of action until the cartridge could be removed. If it happened to fire with the cartridge loaded in reverse, the powder charge exploded with nothing in front of it and just wedged the bullet into the base of the chamber. This front-hinged action, with solid breechblock containing a rimfire or centerfire firing pin, was later used by several countries for converting muzzle-loaders to cartridge rifles, and the U.S. Springfield cartridge rifles employed the same type of action. (Drawing from Rodolph Schmidt's "Les Armes A Feu Portatives," 1877)

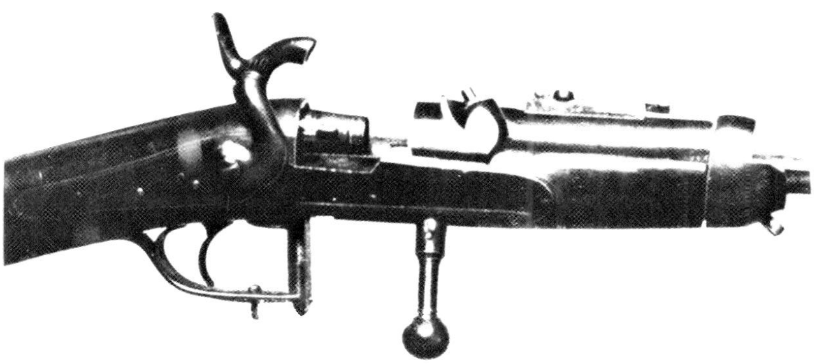

The 1859 English Prince rifle operated on the sliding barrel principle. The operating handle in front of the trigger guard was pushed forward to move the barrel assembly away from the breech. A cartridge was inserted and the barrel drawn back over the extended face of the breech to form a gas seal. (Illustration from "Rifles And Ammunition," H. Ommundson and E. H. Robinson, 1915)

PBL 16 – .56in (14.22mm) U.S. Colt
Revolving Rifle, ca. 1862

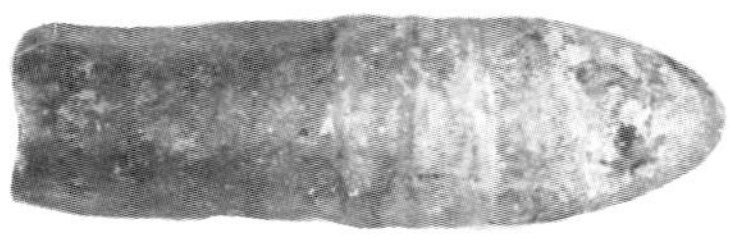

Description

Animal intestine sack filled with black powder cemented to a grooved, conical bullet. Case covered with varnish.

PBL 17 – .56in (14.22mm) U.S. Colt
rifle, ca. 1861-1862
.52in (13.21mm) Sharps
rifle
.52in (13.21mm) Hall
rifle

Black powder pressed in dry form into the shape shown and cemented to the base of a lead bullet. Collodion (nitrocellulose dissolved in ether and alcohol) used to waterproof the powder charge and facilitate ignition.

Comments

The first whole inserted cartridge shown above was made under a Hotchkiss patent prescribing as case material two strips of membrane or gut wound spirally in opposite directions to form a sack for the powder charge. The varnish waterproofed the cartridge and facilitated ignition of the powder by the percussion flame. Another type covered by a different Hotchkiss patent was made with a single piece of animal membrane. D. C. Sage & Co. probably made this specimen which is far less common than the variety of such loads made for revolvers during the Civil War. The one great drawback of all these cartridges was their fragility, which required that they be packed in wooden containers hollowed out to separate each from the other.

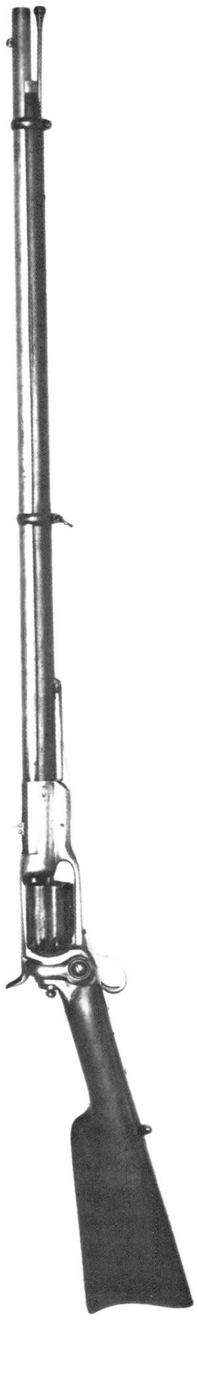

The Colt revolving rifle was actually a chamber loader with a breechblock rotating on a central pin and containing six chambers instead of the one found in those with hinged breechblocks. To operate, each chamber was loaded with paper or combustible cartridge and a percussion cap set on the nipple of each of the chambers. In emergency shortage of ammunition loose powder and bullet could be used as in any of the chamber loaders. Cocking the hammer rotated the cylinder to place a chamber in line with the barrel.

The Colt revolving rifle was made in two frame sizes in calibers .44, .50, and .56, most of them being the .56in. U.S. Ordnance department purchased 666 before the Civil War, the first 101 of which were .44 caliber. During the war a total of 4,172 were bought by the Union.

It has often been written that the Colt rifle was not popular because of its habit of communicating the percussion flame from the cap struck by the hammer to neighboring caps and chambers, resulting in multiple shots at once, one or more of which might strike the shooter's hand supporting the barrel. However, no documented instance of this happening could be found in available sources. Nonetheless, the rifle did not sell as well as others. Some 3100 remained unsold in Colt factory inventory in 1863.

The specimen illustrated here is the Model 1855 rifled musket, caliber .56 (14.25mm) with 37½in barrel.

PBL 18 – .54in (13.72mm) U.S. Greene Underhammer Rifle, 1857

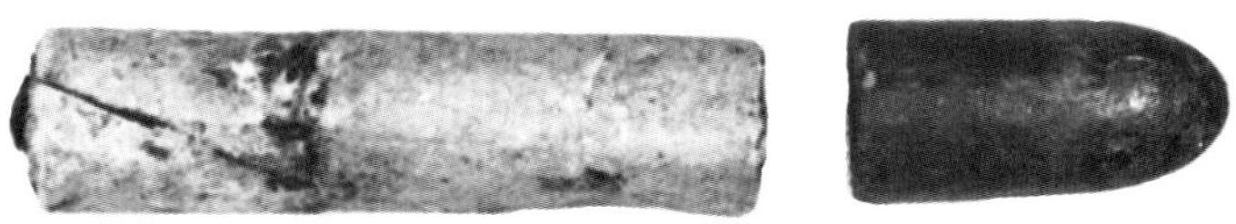

Description

Paper tube loaded with an elongated round-nose bullet with felt wad at its base and powder filled in over the bullet. Case cemented closed over the base of the projectile and also over end of the powder column.

PBL 19 – 13.2mm (.520in) Russian "Obturator" Rifle, 1860

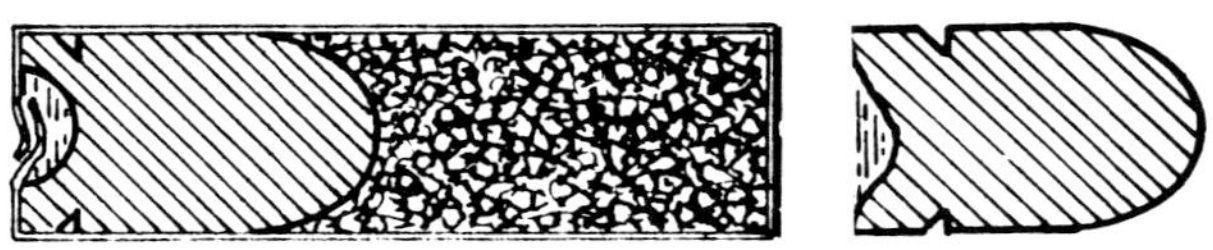

Same construction as above except that the bullet used was of the compression-expansion type with one deep groove around the hollow base. (Drawing by Richard A. King)

Comments

The Green and Russian "Obturator" rifle cartridges were similar in appearance. They are novel in that the bullet (and wad in the Greene) positioned in the rear of the paper case served as the obturating, or gas-sealing, device. The first firing of either rifle required insertion of a separate naked bullet into the throat of the chamber. Then the cartridge was inserted behind this bullet and the bolt closed. The flame from the percussion cap traveled through an ignition orifice which entered the chamber approximately halfway between the two bullets. It penetrated the paper case and exploded the charge. The explosion of the powder forced the rear bullet (also the wad in the Greene) back against the bolt face and prevented the powder gases from leaking out between the face of the bolt and breech. The front bullet, of course, sped down the barrel and toward its intended target. After firing, the soldier manipulated the bolt to shove the rear bullet forward into the throat of the barrel, drew the bolt open, inserted another cartridge behind the bullet which had just acted as the obturating device, closed and locked the bolt, and was ready to fire again.

Lt. Col. J. Durrell Greene, United States Army, patented this underhammer rifle on November 17, 1857. One of the first bolt actions, it was cocked by gripping the ring head of the hammer and pulling down to engage the sear. The sliding cylindrical breechblock was made with an outer tube and an inner bolt. To open, the operating handle was turned counterclockwise to the vertical position and the bolt drawn to the rear. A bullet was pushed into the chamber and the bolt was pushed forward ahead of its locking position. This moved the inner bolt forward seating the bullet in the throat of the chamber. Then the bolt was again withdrawn and a cartridge inserted. The bolt was moved forward only far enough to allow the operating handle to be turned down clockwise to lock the action. This rifle also featured a perfectly smooth interior barrel surface which was bored slightly out of round and given sufficient spiral to function as rifling. Charles Lancaster, the English gunmaker, perfected this system and machinery designed by him was used by Waters of Millbury, Mass. in the manufacture of this rifle. The Union purchased and used 900 of the Greene rifles during the Civil War. Lt. Col. Greene developed and sold an earlier percussion carbine to the U.S. Government based on a different system. Its cartridge is not represented in this study. (Milwaukee Public Museum photo).

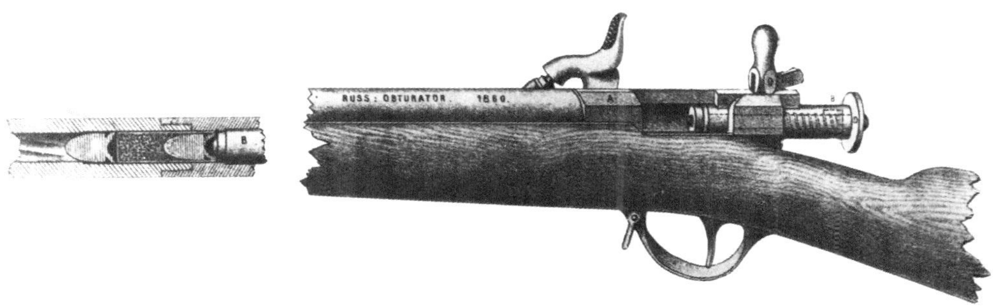

The Russian "Obturator" rifle of 1860 employed essentially the same system as the American Greene. However, the operating handle was turned down to the left in the closed position and the percussion hammer and nipple were in their conventional positions on the right hand side of the breech and chamber. (Drawing from Schmidt's "Les Armes A Feu Portatives," 1877)

PBL 20 – 24 Bore (.577in–14.65mm)
English Calisher & Terry,
ca. 1858

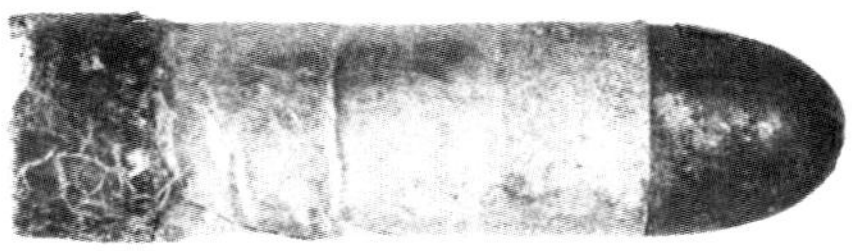

Description

Tan paper cylinder cemented around lower half of a round-nosed, grooveless bullet with a thick, grease-soaked felt wad cemented to the end of the closed powder column.

PBL 21 – 13.9mm (.547in) Saxonian
Cavalry Carbine, 1865

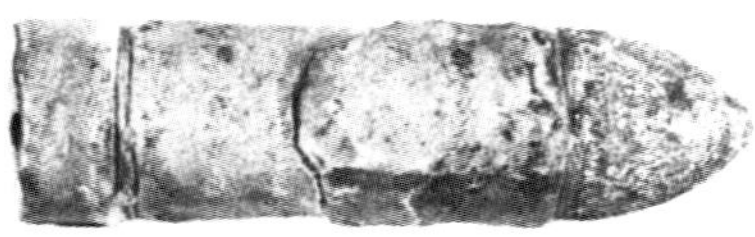

Same general construction as the Calisher & Terry above with blue paper, wad contained within the base of the paper case and secured there with a string tie between it and the powder charge.

Comments

The Calisher & Terry cartridge was the next used in a military percussion breech-loader to include a breech obturating device. The greased wad sat against the face of the breech bolt when the action was closed. The ignition orifice entered the chamber midway between the bullet and the wad. When the powder charge was ignited, the explosive force compressed and expanded the wad against the bolt face and into the wall of the chamber, thus preventing the powder gases from escaping to the rear. When the bolt was withdrawn after firing, the wad remained in the rear of the chamber. The next cartridge pushed it forward into the throat of the chamber and it was ejected ahead of the next fired bullet, at the same time lubricating the bore and taking with it the powder fouling from the previous shot.

PBL 21 was identified when found as for a Calisher & Terry carbine. Such a smaller caliber load may have been used in a model of this arm. However, the bullet diameter and case length of this specimen, as well as the general base construction, corresponds to the Saxonian cartridge illustrated actual size by Thierbach (Fig. 536). The text acknowledges that construction of this cartridge emanated from the English idea developed several years before.

Reference: "Die Geschichtliche Entwicklung der Handfeuerwaffen," M. Thierbach, 1886 (reproduced by Akademische Druck-u. Verlagsanstalt, Graz, Austria, 1965)

The Calisher & Terry of 1858 developed and made in England was also a bolt-action weapon with operating lever hinged to the bolt head. To operate, the hinged lever was lifted out of its locking recess in the receiver and the bolt drawn back. This exposed an opening on the right side of the receiver into which a cartridge was loaded. Confederate forces during the American Civil War used this arm in the carbine form shown below. The felt wad obturating device evidently functioned as intended and its lubricating action as it left the bore ahead of the next bullet fired softened or eliminated powder fouling residue.

No illustration of the Saxonian cavalry carbine could be found but Thierbach shows it as a sliding bolt type action somewhat similar to the Calisher & Terry. (Milwaukee Public Museum photos)

PBL 22 – .577in (14.65mm) English
Wilson Rifle, ca. 1860

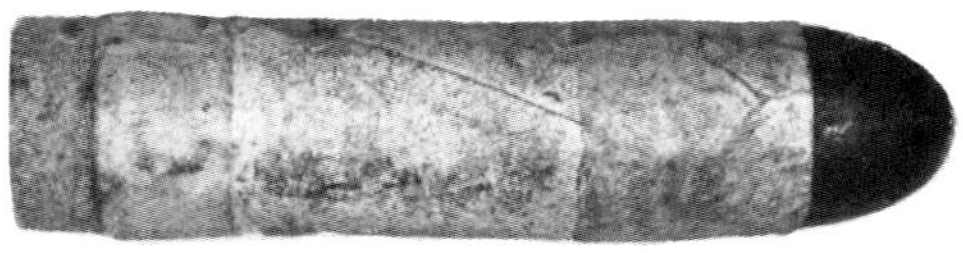

Description

White paper cylinder cemented around two-thirds of the length of an ungrooved round-nosed bullet. Lubricated felt wad attached to the base of the powder column with an outer wrapping of paper.

PBL 23 – .442in (11.23mm) English
Westley Richards "Monkey Tail"
Carbine, 1862

Brown paper cylinder similar to above with bullet separated from the powder charge by a string tie, and greased wad cemented externally to the end of the powder column.

Comments

These two cartridges functioned in the same manner as the Calisher & Terry and Greene types. All of the English cartridges made in this fashion were wrapped in protective paper envelopes with a fabric tear strip for opening. Because of this, they are difficult to identify. However, the envelopes are easily opened with little damage if some care is exercised.

The 1860 Wilson rifle, made in England, could be called a bolt-action arm. To unlock and open the action, a flattened key, operating through a slot in the right side of the receiver wall, was pulled out to disengage it from the side of the bolt. The bolt could then be drawn back by the hinged piece on its head. A cartridge was inserted in the chamber and the reverse procedure closed and locked the arm for firing. As in the Terry, the lubricating base wad functioned to seal in the exploding powder gases. (Milwaukee Public Museum photo).

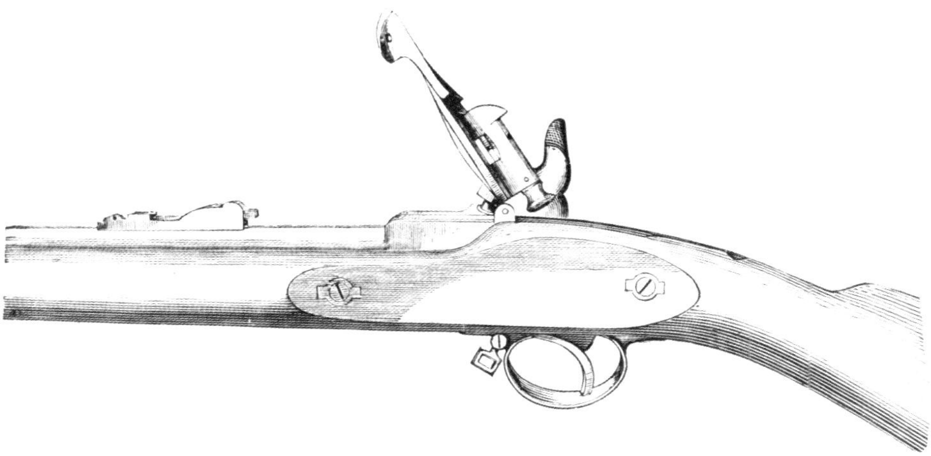

The Westley Richards "Monkey Tail" carbine, so named for the appearance of its uplifted lever and plunger bolt, was a camming breech bolt arm. When the operating lever was raised, the bolt head moved in an arc up out of its recess behind the chamber and back down in reverse arc when the lever was lowered. The closing action forced the cartridge tightly into the chamber. The percussion flame ignited the powder between the rear wad and the bullet. This carbine saw service in the British Army Yeomanry between 1862 and 1881, and was also used by Colonial forces during the Maori Wars in New Zealand. A rifle version was issued experimentally to several British Army battalions but never officially adopted. (Drawing from W. W. Greener's "The Gun and Its Development," 1899)

References: "Early Percussion Firearms," Lewis Winant, 1959
"19th Century Firearms," Maj. F. Myatt, MC, 1979

PBL 24 – .54in (13.72mm) U.S. Marston Rifle, ca. 1853

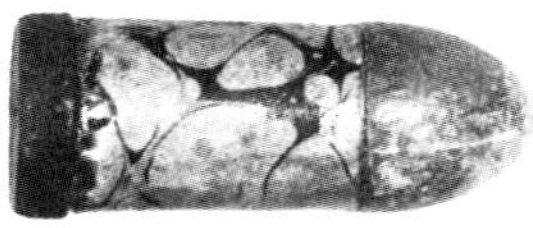

Description

Cardboard tube of mixed mottled colors with bullet seated friction tight in one end, powder filled into remainder, and folded to center but not closed. A greased leather washer with center perforation is cemented to the base.

PBL 25 – .36in (.9.14mm) U.S. Marston Rifle, ca. 1853

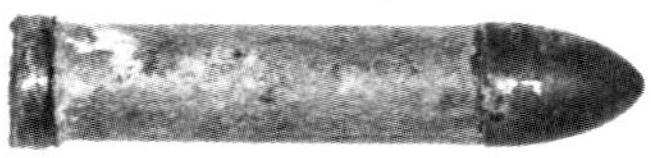

Same construction as above except for color of case tube which is blue in this specimen.

Comments

The Marston cartridges and arms were developed some years prior to the English Calisher & Terry, Wilson, and Westley Richards type. It is known that the Marston system was tested in England about 1853 and received favorable reports. Thus, it seems likely that the idea contributed to the English developments. Since ignition occurred through the base perforation, the case could be made of stiffer, more durable material. Even if this did not burn completely, the washer which remained in the chamber after firing pushed any residue ahead of it and out the barrel when the next cartridge was loaded and fired.

The Marston cartridges were made in a number of rifle, pistol, and shotgun sizes but never saw military service, probably because of lack of power. They are included here because they represent a significant step in solving the problem of gas escape at the breech. The leather washer evidently expanded effectively against the bolt face and chamber wall to contain all the explosive force of the powder charge for forward thrust.

W. W. Marston's action patented June 18, 1850 in the United States never gained the attention of military authorities. However, the principle of igniting the powder charge through a central hole in the base of the cartridge was a significant development adapted to several successful American percussion breech-loaders used during the Civil War—namely, the Maynard, Gallager, Burnside, and Smith.

To operate, the trigger guard lever was pushed down and forward to draw the sliding bolt to the rear, exposing the loading port in the side of the receiver. After dropping in a cartridge, reversing the procedure moved the bolt forward to cam the cartridge into the chamber. The wad from the previous shot, plus any case or powder residue, was pushed into the barrel by the fresh cartridge. The wad traveling ahead of the bullet cleaned the bore. (Milwaukee Public Museum photo)

Description

PBL 26 – 14.5mm (.570in) Bavarian Lindner Rifle, 1867

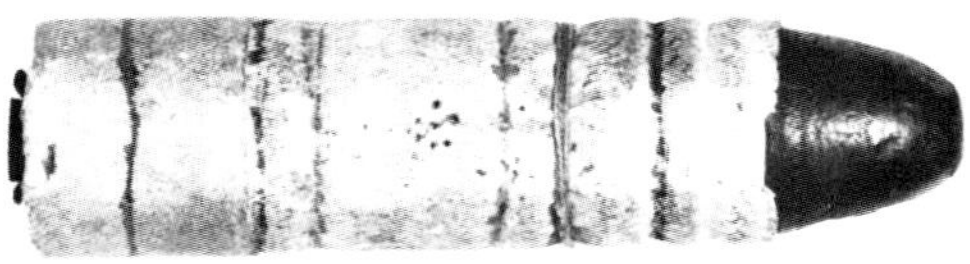

White paper tube wrapped around a two-grooved bullet with a thread tie just below the bullet. The powder column is closed with a papier-mache wad over which the case is folded and an extra wrap of paper cemented over the base exterior. The base wad is formed with a cavity deep enough to hold an inverted percussion cap.

PBL 27 – 14.5mm (.570in) Bavarian Lindner Carbine, 1867

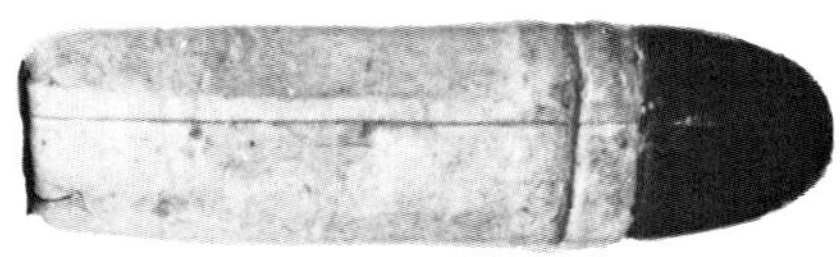

Same construction as above with shorter case and no extra wrap of paper over base.

PBL 28 – 14.5mm (.570in) Bavarian Lindner Explosive

Same construction as above but no thread tie below bullet. The "Podewil" explosive bullet is shown sectioned (percussion cap fitted over a tube filled with detonating powder and inserted in cavity), whole bullet, and bullet recovered after firing.

Comments

The Bavarian Lindner cartridges are found with a number of small pinholes (specifications called for seven) midway in the side of the paper case. These were intended to ensure penetration of the case by the percussion flame should the double fold of paper be under the orifice. The papier-mache wad in the base evidently served as a breech obturating device. The explosive force of the charge compressed it into the face of the bolt and chamber wall as in the Marston, Greene, and English types previously described. The rifle had no extraction mechanism, so the wad remained in the chamber and was pushed forward by the next cartridge loaded.

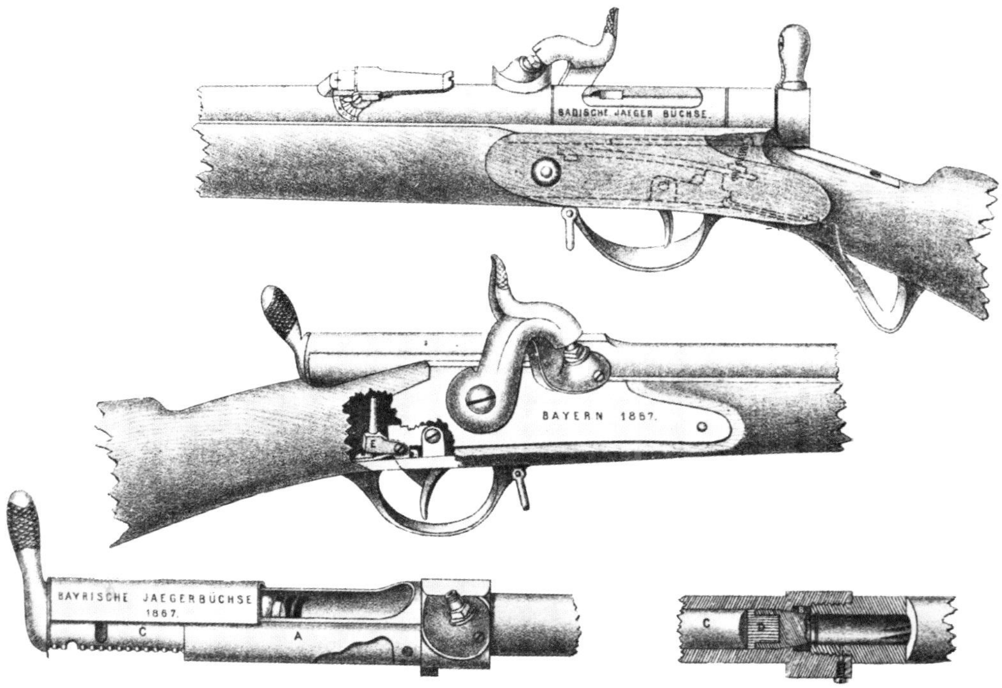

The Bavarian percussion breech-loader of 1867, developed by Edward Lindner (center illustration), was a bolt-action arm with ignition orifice in the center of the cartridge. The related Badische Jaeger rifle shown above had a different trigger arrangement and the nicety of a metal pistol grip. The lower illustration reflects the locking threads on the bolt body of these rifles and an obturating device on the face of the bolt to prevent gas escape to the rear. (Drawing from Schmidt's "Les Armes A Feu Portatives," 1877)

PBL 29 – .50in (12.70mm) U.S. Smith Carbine, 1857

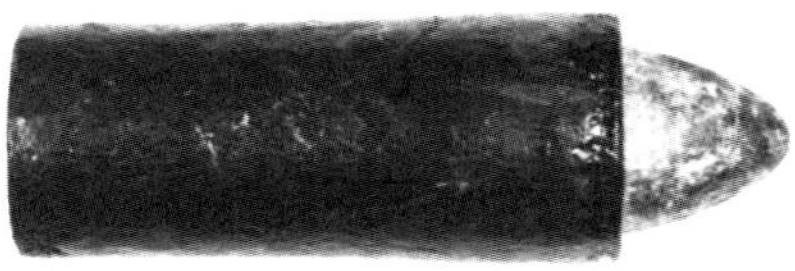

Description

Black India rubber cylinder with conical bullet fitted friction tight in mouth. Base of powder chamber closed with a disc of the same material, perforated in the center, and cemented in place.

PBL 30 – .50in (12.70mm) U.S. Smith Carbine, 1863

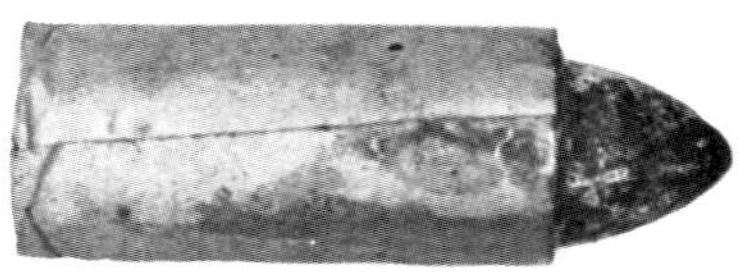

White paper covering pasted over a cylinder constructed of wrapped brass foil folded at right angles to the case wall to close the base, leaving a central ignition orifice. Powder occupies one half of the case and one-groove lead bullet the remainder.

Comments

The rubber case carbine cartridge patented by Gilbert Smith in June 1857 provided another solution to breech obturation. The flexible case material expanded tightly against the chamber walls when the powder exploded and then contracted after pressure and temperature diminished so that it could be removed from the chamber with the fingers. These rubber cases could be reloaded but percussion flame and powder gases eroded the ignition orifice as well as the case walls. Accordingly, reloadability was not an important feature. However, specimens exist with a brass or copper base disc which could withstand repeated firings.

The paper-covered foil case was made under the December 1863 patent awarded to Thomas J. Rodman and Silas Crispin, which was assigned to Thomas Poultney of Baltimore Maryland. This also was an effective breech-sealing system. The wrapped brass case was far more durable than the rubber type, although it is doubtful if they were reloaded. Over 13,000,000 of these cartridges were made during the Civil War years and a lot of them survived judging from their presence in most U.S. cartridge collections. Collectors find the diagonal seam paper covering over the case much less common than the straight seam. The rubber case type is a scarce item. It also exists in .36 caliber.

The Smith carbine was operated by depressing the lever ahead of the trigger to release the top-strap latch that held the barrel group to the breech face. The barrel then pivoted on the pin ahead of the trigger guard. The chamber was half in the barrel and half in the face of the breech. With cartridge inserted in the barrel, the gun was pivoted closed. After firing, the empty case was held either by the base or mouth so that it could be grasped with the fingers and extracted. Either type cartridge used in this arm effected a complete gas seal.

The Smith carbines were made by the Massachusetts Arms Company, American Machine Works, Springfield, Mass., and American Arms Company of Chicopee Falls, Mass. for Poultney & Trimble, Baltimore, Md., the company that held the contract for them with the U.S. Government. (Milwaukee Public Museum photo).

PBL 31 – .54in (13.72mm) U.S. Burnside Carbine, Foster Improvement of April 1860

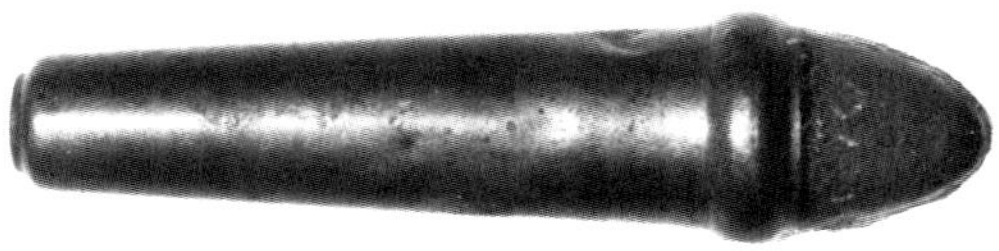

Description

Case of drawn brass tapering to the rear with base ignition orifice. Base is slightly extended and flattened. The annular swell at the case mouth around the bullet is filled with lubricant.

PBL 31(b) – .54in (13.72mm) U.S. Burnside Carbine, common form

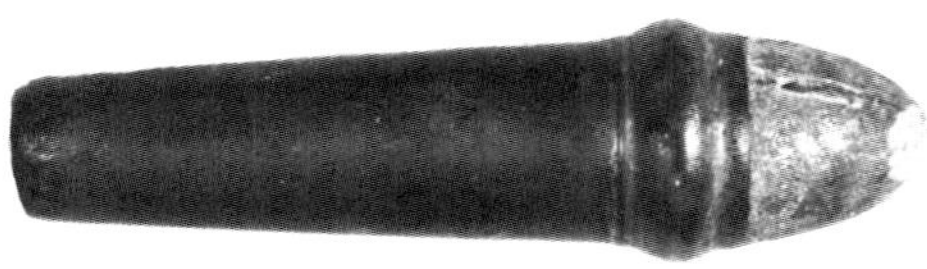

Same as above but shorter case and rounded perforated base.

PBL 32 – .54in (13.72mm) U.S. Burnside Carbine, Rodman-Crispin patent, 1863

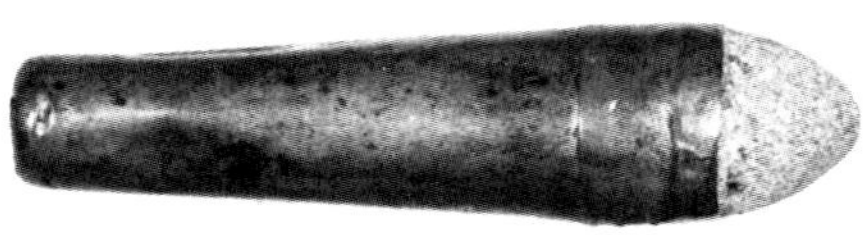

Case of wrapped brass foil with diagonal seam tapering toward an ignition orifice in the base. Three-fourths of the case filled with black powder and a round-nose bullet inserted friction tight in the case mouth. No lubrication around bullet.

Comments

The Burnside patent drawing, opposite, shows a chamber and cartridge with no grease-containing swell around the bullet. No specimen of this first form of the cartridge made as early as 1856 could be found to illustrate. The first and second specimens shown above represent the earliest mass-produced cartridge for military service using a drawn brass case. Solid metal provided a perfect gas seal and the Foster improvement provided lubricant to keep the bore swabbed clean with each shot.

The Poultney patent cartridge, made without provision for the lubricant, is far less common than the drawn brass variety. Drawn brass specimens will be found with a dished base around the ignition orifice, and a tinned brass variant also exists. Both are in the very scarce category. Dished base fakes have been created from the most common form at PBL 31(b), so the collector should exercise care in buying or trading for a specimen.

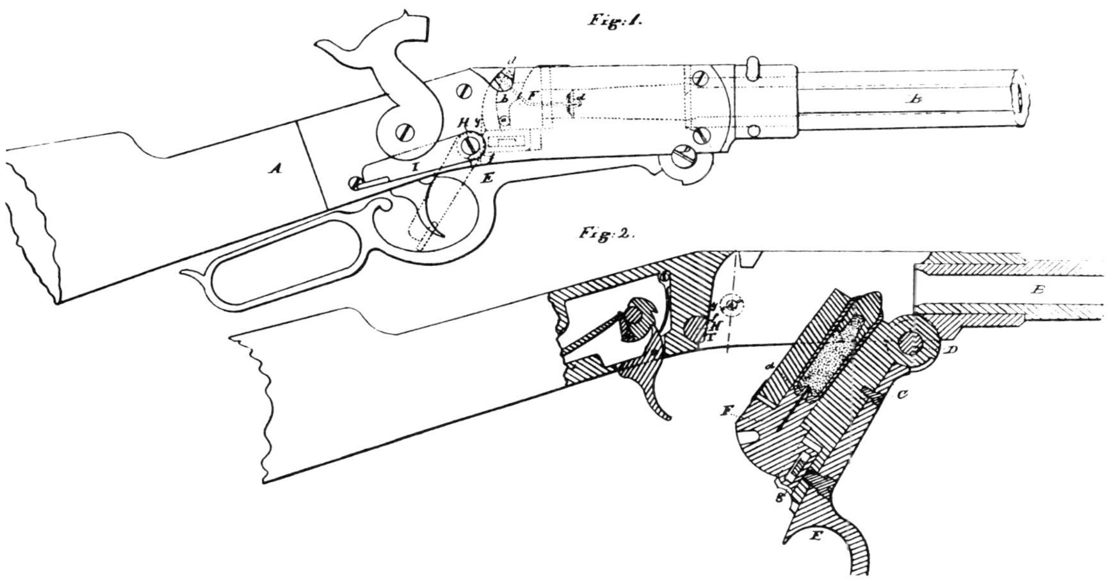

This breech-loading carbine, patented by Ambrose Burnside in March 1856 and later improved by application of the Hartshorn and Foster patents, was operated by dropping the trigger guard underlever which pivoted the breechblock down and away from the barrel. The reverse taper cartridge was dropped into the chambered breechblock and the underlever brought up to bring the chamber in line with the barrel. Only the bullet and mouth of the case protruded into the barrel. The solid drawn or foil brass case expanded at the moment of discharge to seal the breech/barrel joint. The Union purchased over 55,000 of these arms during the Civil War. The upper illustration, from an original patent drawing, shows the first configuration of the Burnside. The photograph is of the second model with the Foster improvement patented April 10, 1860 incorporating a rocking curved lever inside the trigger guard/operating lever which functioned to unlatch the action when pressed forward. Later models made during the Civil War had a wooden forend. Burnside Rifle Company, Providence, R.I. manufactured this carbine, though Bristol Firearms Co. made some of the second model. (Milwaukee Public Museum photo)

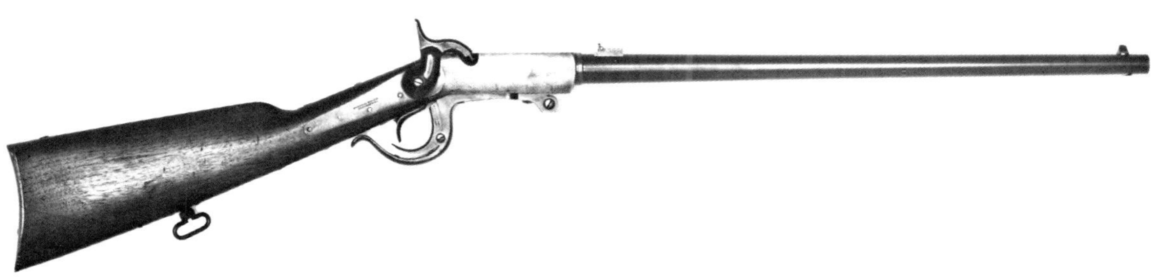

PBL 33 – .50in (12.70mm) U.S. Maynard Carbine, 1859

Description

Brass tube folded over at base with a central orifice remaining, soldered to a thin steel or brass base disc. A flat-nose conical bullet seated friction tight in the case mouth atop the powder charge. No wad between powder and bullet.

PBL 34 – .50in (12.70mm) U.S. Maynard, ca. 1863

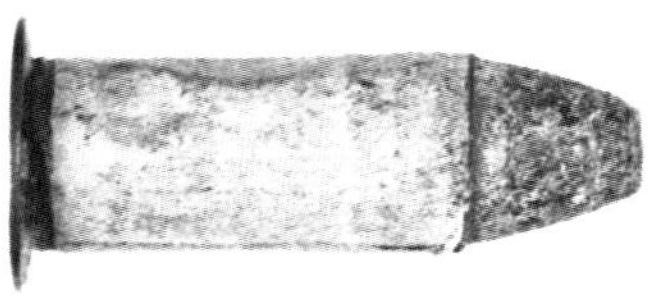

Brass foil tube covered with white or beige paper soldered to a brass base disc. Central base perforation as in first type.

Comments

Dr. Edward Maynard's patent of June 17, 1856 illustrates a perforated base cartridge with no rim. However, the relatively common form of the .50 caliber Maynard percussion cartridge for military carbine is the drawn brass tube form with brass base disc forming a wide rim. Some of the earliest of these had a steel base disc. Other cases were tinned and shorter bullet loads also exist.

The second type was made under the patent taken out in December 1863 by Silas Crispin with Thomas Rodman, which was later assigned to Thomas Poultney. Specimens exist with straight seam on the paper cover as well as diagonal. Both are uncommon.

Why the makers of this ammunition used brass foil with paper covering after the drawn brass tube had been proved successful is not clear. However, judging from the number of them that survived the Civil War, the drawn brass tube must have been produced throughout that conflict.

Maynard percussion cartridges made in the form of PBL 33 exist in .35, .40, .55, and .64 calibers for sporting rifles and shotguns.

Dr. Edward Maynard patented this underlever, hinged-barrel carbine in December 1859. Rotating the trigger guard forward allowed the barrel to drop by its own weight. The closing action cammed the standing breech and barrel face tightly together. The drawn brass or Poultney paper-covered foil case sealed the breech against escape of powder gases. This arm could be taken down by giving the crosspin on the right side of the receiver a half turn and pulling it out of the receiver. This separated the operating lever group from the receiver and the barrel then lifted off the hinge pin. Most of these carbines were not equipped with Dr. Maynard's tape primer since he sold the rights to the patent to the U.S. Government before he developed this weapon. Massachusetts Arms Company made and sold over 20,000 of the Maynards to the Union during the Civil War.

After the war, this system formed the basis for a successful line of sporting rifles and shotguns, and was ultimately adapted to self-contained centerfire cartridges. (Milwaukee Public Museum photo)

PBL 35 – .50in (12.70mm) U.S. Gallager Carbine, 1860

Description

Drawn brass case with an ignition orifice in the rounded base. A conical bullet is seated friction tight in the case mouth atop the powder charge.

PBL 36 – .50in (12.70mm) U.S. Gallager Carbine, ca. 1863-1865

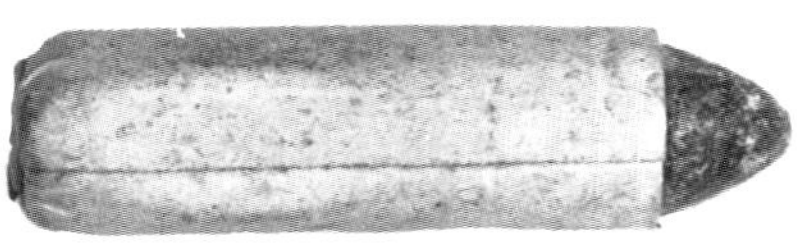

White paper covering on a Poultney-patent wrapped brass foil case with ignition orifice in base as above. Specimens also exist bearing paper covering with a diagonal seam.

Comments

The Gallager carbine cartridge has been found in .58 caliber as well. One specimen has a solid drawn steel case, another with case of metal other than steel or brass. The drawn brass .50 caliber type was made with a paper lining; the Jackson's patent case was constructed of tinned iron covered with paper detectable with a magnet; another variant found has a soldered seam on the foil case with no paper covering. Another was made with tinned brass case, and a linen patched ball is also known. PBL 35 exists with "Patented" stamped around the base perforation. PBL 36 is the most commonly found specimen among the Gallagers.

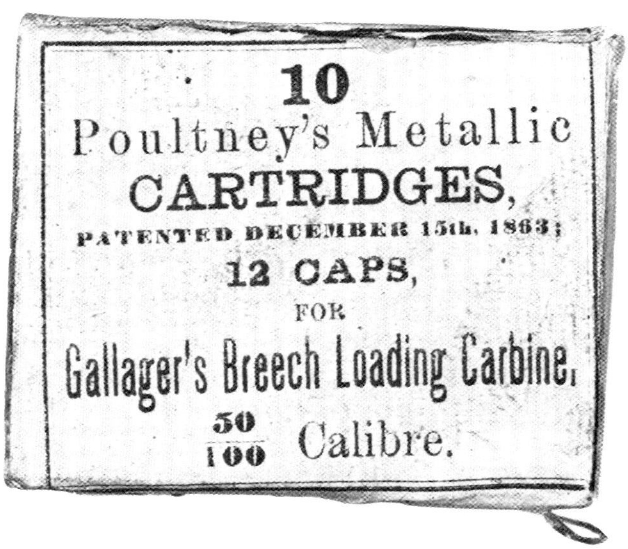

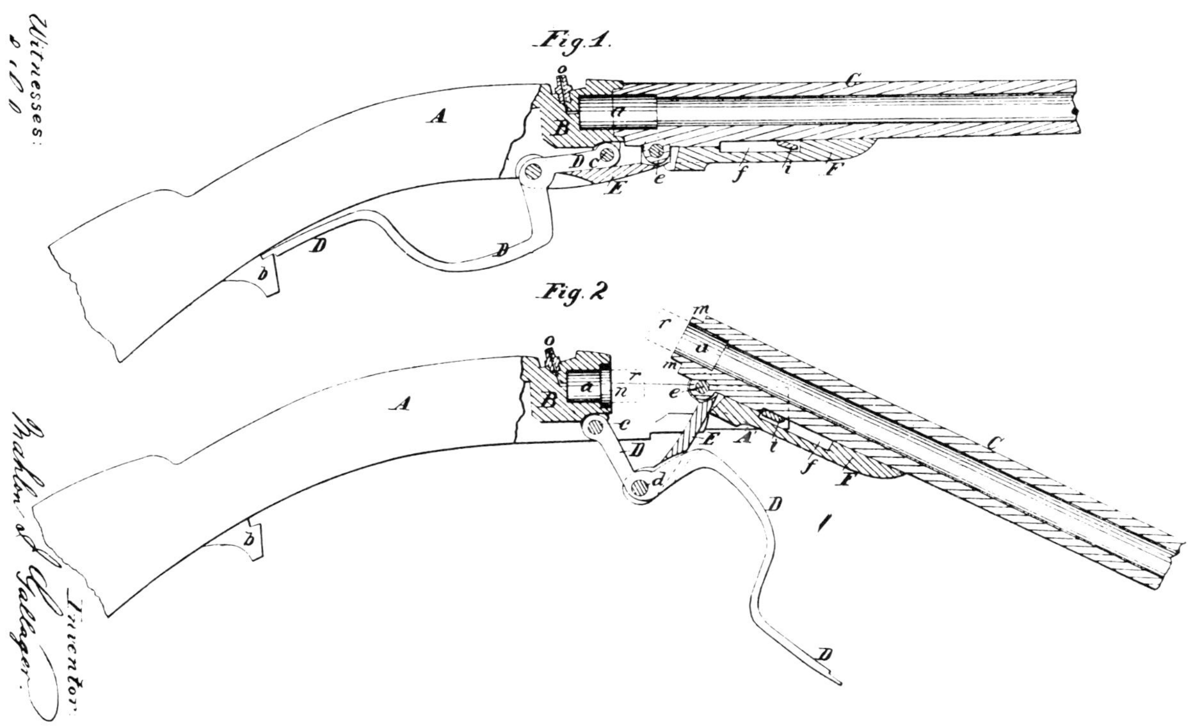

Mahlon J. Gallager's carbine patented in July 1860 operated by pushing forward the trigger guard underlever which moved the barrel forward and let it drop by its own weight to expose the chamber. As in the Smith carbine, half of the chamber was in the rear of the barrel, half in the face of the breech. The empty case remained either in the breech end or the barrel end so that it could be extracted between thumb and forefinger.

The Gallager carbines were manufactured by Richardson & Overman of Philadelphia, Pa., who sold the Union 22,278 of them during the war. (Illustration from the original patent drawing)

PBL 37 – .69in (17.53mm) U.S. Maynard "Capsule," 1857

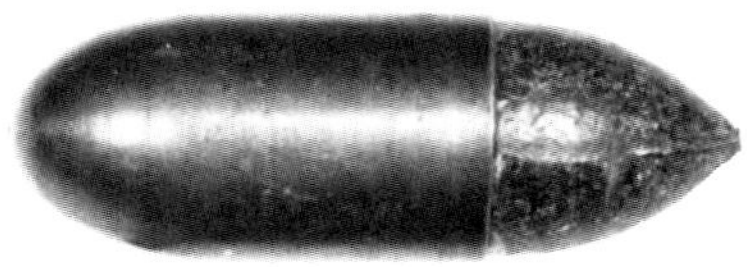

Description

Drawn brass case with rounded perforated base and conical bullet seated friction tight on black powder charge.

PBL 38 – .69in (17.53mm) U.S. Maynard "Flop Ear," 1857

Drawn brass cylinder with separate brass perforated base disc made with a manual extraction handle. Base disc soldered to case body.

PBL 39 – .69in (17.53mm) U.S. Maynard Cord Extractor, 1857

Drawn brass cylinder with base disc, same as above except for length of cord attached to base which first served to carry the percussion cap, and then as an extraction device.

Comments

These cartridges were designed by Dr. Edward Maynard for use in the rifle actions of his design for converting muzzle-loaders to breech-loaders. He entered these systems in the U.S. 1857 trials where they were unsuccessful. The capsule type evidently functioned in an action where the rear part of the cartridge was seated in the breech face (similar to the Smith and Gallager carbines), and the forward portion in the rifle's chamber. The lever and cord extractor types had to be positioned in the chamber so that the appendage for extraction protruded from the action through a mortise.

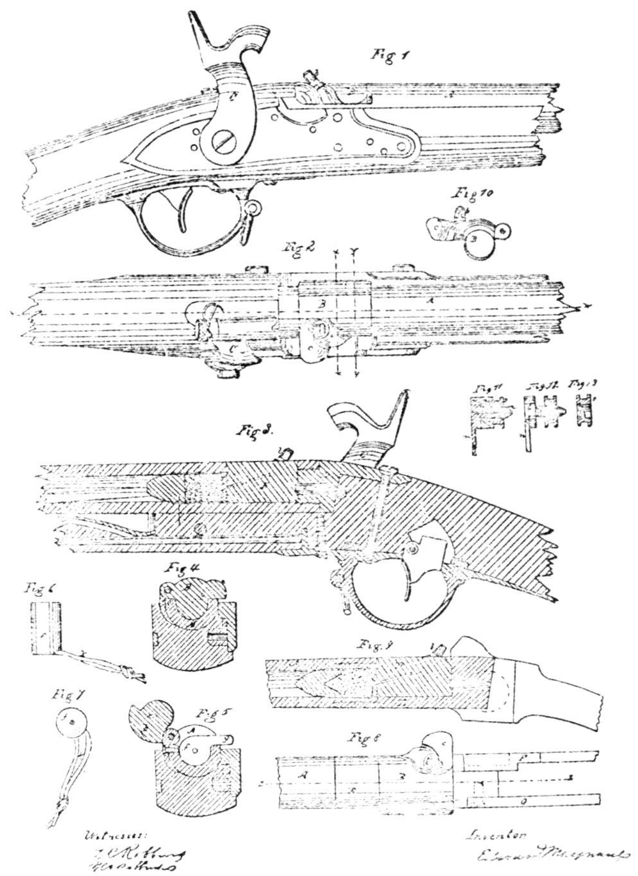

One of Dr. Edward Maynard's actions designed for converting muzzle-loading rifle to percussion breech-loader using cord (also thong) or arm extraction cartridge shown opposite. The breechblock hinged at the left swung counterclockwise to expose the chamber for loading. It is evident from the variety of manual extraction devices he proposed that the idea of mechanical extraction of an empty cartridge case had not occurred to him at this stage. (From an original patent drawing)

Dr. Maynard also produced this type of cartridge in .58 caliber with arm extractor and in .50 caliber with cord or wire extractor. The Smithsonian Institution collection of Maynard patent items includes .69 caliber specimens with longer case to contain ball and shot loads. Collectors should demand a guarantee of authenticity before trading for or investing in one of these experimentals. Very few were made, but more than a few duplicates have been created as the interest in these has grown. Some exist in sizes which would fit no known firearm; i.e., .75 caliber with arm extractor.

PBL 40 – .54in (13.72mm) Hunt "Rocket Ball,"
U.S., 1848

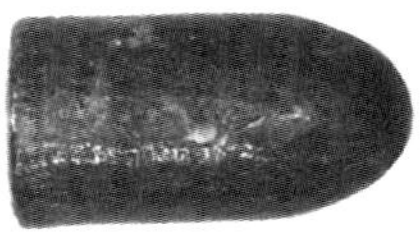

Description

Hollow lead conical bullet filled with black powder and closed with a centrally-perforated cork wad. Stamped on the side–"Patented 1848." Specimens exist with the patent stamp running around the circumference of the bullet, and with no patent stamp. A shorter .89in variant exists, both stamped and unstamped.

Comments

The Hunt cartridge was used in the original Hunt tubular magazine rifle as well as the subsequent improvements of the arm made by Lewis Jennings. This was the only percussion breech-loader designed to be fed from a separate magazine to the chamber by mechanical means. If formed the basis for the idea subsequently developed by Daniel Wesson and Horace Smith of a powder charged conical lead bullet containing its own priming which was used in the Volcanic pistols, carbines, and rifles.

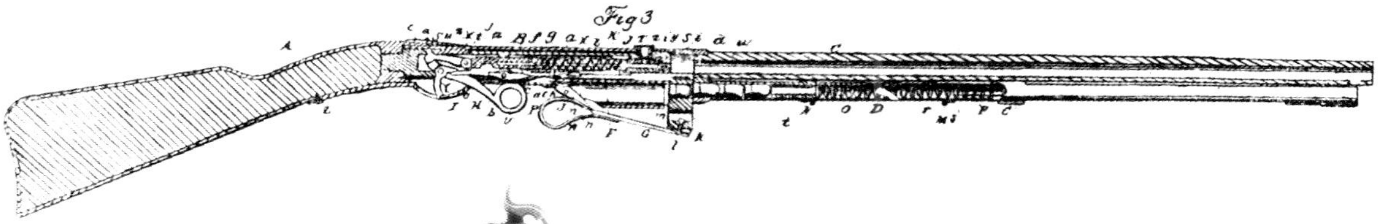

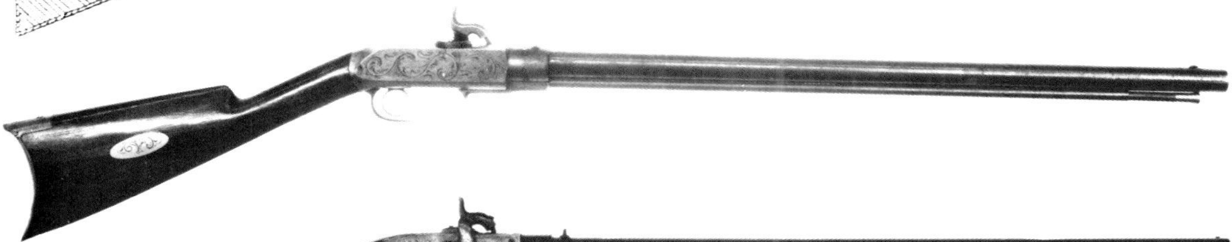

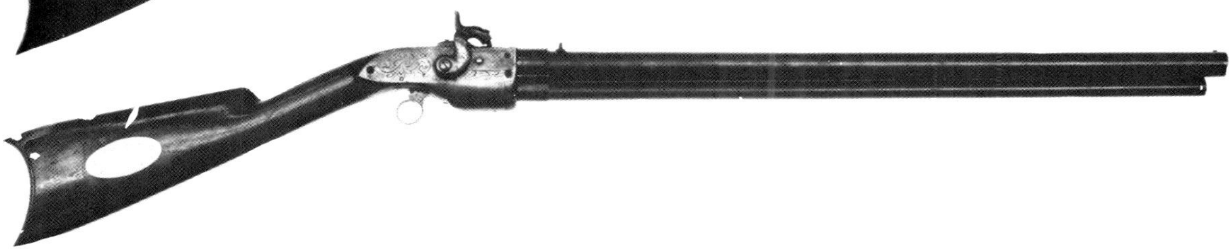

Walter Hunt patented the upper illustrated firearm in the United States on August 21, 1849 and called it the "Volition Repeater." He evidently had no funds to produced it and assigned the patent to New York model maker and machinist, George Arrowsmith. Arrowsmith had in his employ one Lewis Jennings, also an inventor. They evidently decided the Hunt rifle was too complicated to produce and Jennings proceeded to modify it. On December 25, 1849, he was granted a patent on his improvements of the lock and repeating mechanisms. Arrowsmith subsequently sold the Hunt as well as the Jennings patents to Courtland C. Palmer who contracted with Robbins and Lawrence of Windsor, Vermont to manufacture the Jennings model. However, the repeating mechanism was still too complicated to work reliably and most of the early Jennings rifles were finished up as single shots featuring a sliding ring trigger arrangement within the guard which drew back the breech bolt when it was pushed forward and closed and fired the action when it was drawn back to the rear (second rifle illustrated above). Horace Smith, later of Smith & Wesson, was associated with Palmer and further experimented with the Hunt/Jennings action. He took out a patent on his own improvements on August 26, 1851, reflected in the lower rifle shown above. This arm, with tubular repeating mechanism, was made in some quantity. All three guns used the .54in Hunt rocket ball. Ignition was accomplished by Maynard tape primer fed from a magazine located in the top of the frame. The cartridge had insufficient power to make the Smith-improved Jennings a success and the whole idea was abandoned in 1852.

The Hunt and improvements on it introduced to the arms world the idea of a tubular, under-barrel magazine. The Volcanic, Henry, and Winchester lever-action arms evolved from this idea, as well as the later sporting rifles developed or made by Burgess, Marlin, Colt and others. U.S. military authorities ignored the idea but Switzerland, Austria, Germany, Sweden, France, and Japan developed tubular magazine military rifles, all of which descended from Walter Hunt's 1849 patent. (Drawing after the original patent illustration. Milwaukee Public Museum photos)

PBL 41 -- .56in (14.22mm) Billinghurst Requa Mitrailleuse, U.S., ca. 1861-1865

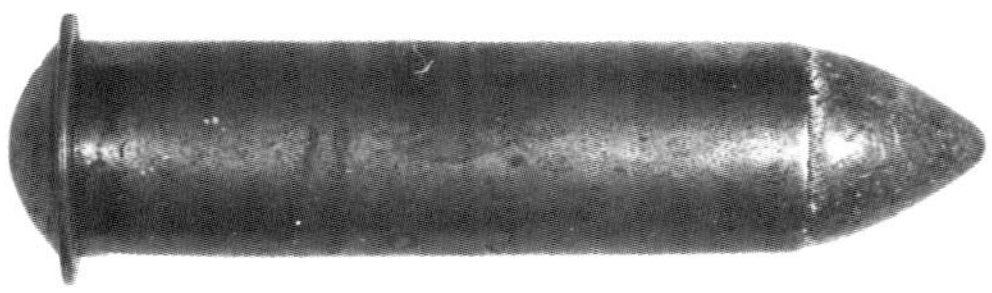

Description

Brass cylinder closed at one end with a soldered on head forming a wide rim and bulbous base which is perforated in the center. A conical bullet is seated in the case mouth atop the powder charge.

Comments

The Billinghurst cartridge fired in the 25-barrel volley, or battery, gun shown opposite was one of the first to use an all-metal case. Fuller, quoting verbatim from an original 1861 promotional pamphlet concerning the gun, describes the case as being made of copper. Copper specimens may exist, but the brass type is the one seen in cartridge collections. The cartridges were loaded with loose powder and conical bullet or, according to the promotional description, with a patented skin cartridge, evidently similar to PBL 16 (p. 68). The loaded rounds were inserted in a row of holes in a long, hollow strip, or sliding bar, presumably of metal. This device spaced them to allow their insertion all at once in the 25 chambers. Used during the Civil War to a limited extent, a substantial quantity must have remained unfired since it is, in brass case form, relatively common in the United States. A puzzling variation is the occasional specimen that got by the inspection stage during manufacture with no ignition perforation in the base.

The 25-barrel Billinghurst Requa Mitrailleuse (a battery gun) was an advancement of the principle shown on page 27 (Danish orgelespingol). A breech bar extending across the rear faces of the barrels was hinged at each end so it could be tipped open or closed by means of an upright lever at each end of the breech.

To operate, the perforated strip loaded with 25 cartridges was used to insert the rounds in the chambers all at one time. A thin train of black powder was laid along the exposed perforated bases of the cartridges and the breech bar levered closed. The hammer on a centrally-situated lock was cocked and a percussion cap fitted on the nipple. The gun was fired by pulling a lanyard. The single percussion cap exploded by the falling hammer ignited the powder train and set off all of the cartridges simultaneously.

This gun was used to some extent by Union forces during the American Civil War for bridge defense. Confederate forces used the single-barrel 1.75in (44.45mm) Williams machinegun in several battles. This crank-operated weapon fired paper combustible cartridges at the rate of 65 shots per minute. No specimen of the cartridge could be located for this study. It actually would fall in the category of small cannon rather than small arm.

Reference: "Small Arms of the World," 5th ed., 1955, W.H.B. Smith (West Point Museum photo)

PBL 42 – .58in (14.73mm) U.S. Ager (also Union) Machinegun, 1861

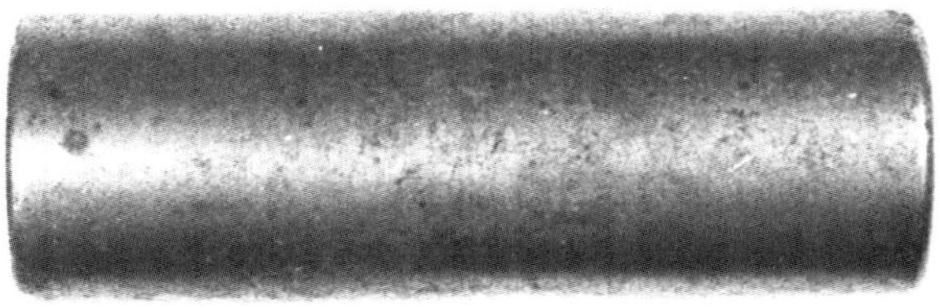

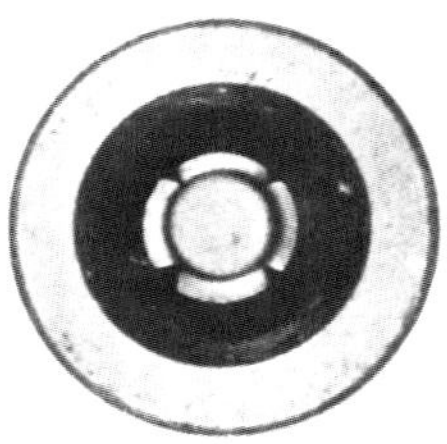

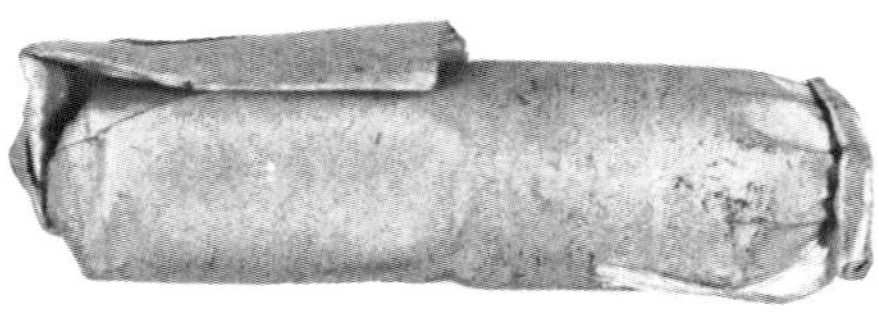

Description

Hollow steel cylinder closed at one end and threaded to receive a steel percussion nipple with its upper end approximately flush with that end of the cylinder. A percussion cap is shown fitted on the nipple in the center illustration. Below that is shown the paper-case, muzzle-loading .58 caliber cartridge used to load the cylinder.

Comments

The steel cylinder used in the Ager (Union Repeating Gun) was not a cartridge, but a removable breech piece reminiscent of such devices used in the earliest breech-loading firearms. Theirbach discusses and illustrates several of these, the earliest of which he dates 1340 (Fig. 460). However, it is more accurate to call the device used in the Ager an auxiliary chamber. Loaded with the paper cartridge shown below it and primed with a percussion cap, it constituted ammunition. A similar cylindrical auxiliary, some three-eighths of an inch shorter, was used in the first Gatling gun. A supply of these cylinders, the cartridges, and primers were carried in ammunition chests on either side of the gun upon the carriage. The cylinders ejected upon firing were recovered and reloaded.

Reference: "Die Geschichtliche Entwicklung der Handfeuerwaffen," M. Thierbach, 1886 (reprinted by Akademische Druck- u. Verlagsanstalt, Graz, Austria, 1965)

This primitive machinegun was apparently developed around 1861 in the United States but no patent was taken out for it until 1866 when Wilson Alger, an American, secured British patents for such a gun. During the American Civil War it was known as the Union Repeating Gun and the Coffee Mill Gun, due to the appearance of the cartridge feed hopper on top of the breech.

To operate, the hopper was loaded with the steel cylinders containing musket cartridges and primed with percussion caps. Working a hand crank pushed one cartridge forward and held it in alignment with the barrel as a wedge lock was cammed forward to fire. Rate of fire was 120 rounds per minute.

This gun was experimented with to some extent during the Civil War but its barrel heated badly and it gained a reputation for breaking down. The Gatling gun was patterned after this mechanism, substituting a hand-cranked multi-barrel group which revolved to accomplish the loading, firing, and ejecting stages of operation. Like the Alger, it was not successful with steel loaded cylinders for ammunition. Constructed to use rimfire cartridges, and later centerfire ammunition, it was a world-wide success (See Chapter 11, "The Rimfire.")

Reference: "The Gatling Gun," 1965, Paul Wahl and Don Toppel
"Small Arms of the World," 1955, W. H. B. Smith

PBL 43 – .56in (14.23mm) Unknown, English Origin, 1869

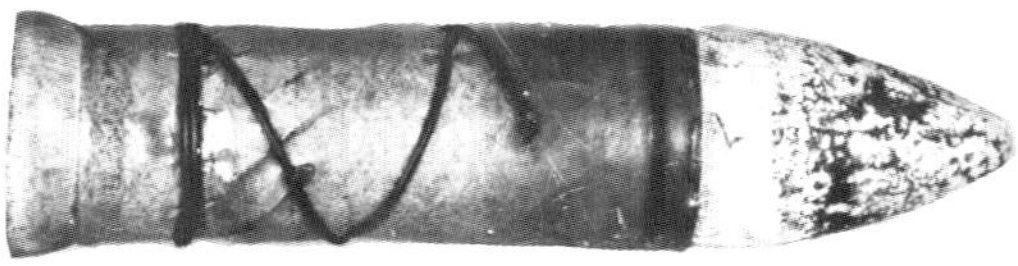

Description

Light brown paper case loaded with a conical lead bullet, and powder column closed with a tapered plug of wood. A strand of copper wire is tied around the case circumference near the base, wrapped around it toward the base of the bullet, with the other end apparently attached to the bullet or wad through the paper case.

Comments

Bartlett & Gallatin's "Cartridge Manual" listing cartridge patents issued in the United States, England, and France prior to 1878 shows a similar specimen on Plate 11 of the English section–2071 of 1869. The description reads–"Case of paper, with metal thimble at base, and rubber wad between. Wire from front wad carries out paper case when fired."

This is a variation on the same idea patented in England by Manceaux in 1859, though Rudolf Schmidt dates it 1861 (See PBL 13, p. 64). In this English specimen, the wire would tighten around the unburned paper case material as the bullet moved forward. The wad apparently remained at the base of the chamber against the breech face.

The arm which used this cartridge is unknown. However, the English modified some Model 1855 caliber .56 Sharps carbines by removing the tape priming mechanism and situating a percussion nipple on top of the chamber so that the ignition flame from the percussion cap would enter the center of the chamber in the approximate middle of the paper cartridge case. It is conceivable that the cartridge shown here was tried out in such a rifle. The wood plug, a variation from the patent description, would be forced tightly against the wall of the chamber and against the face of the sliding breechblock when the powder charge exploded, thus providing a gas seal. When the gun was opened, the tapered plug would fall out or could have been picked out of the chamber. Although it is not apparent, a rubber wad as described in the patent summary may be inside the base in front of the wooden plug.

Reference: "Sharps Firearms," Frank Sellers, 1978, p. 57 (illustration of modified Sharps 1855 carbine)

PERCUSSION BREECH-LOADING ARMS AND CARTRIDGE DATA

No.	Bore	Model and Date	Bullet Diameter	Bullet Weight	Powder Charge	Case Length	Cartridge Length	Sources
PBL 1	.64in 16.25mm	U.S. Hall Carbine, Model 1839	.640in 25.76gm	397.5grs 4.86gm	75grs		1.75in 44.45mm	Specimen—Lewis
PBL 2	.52in 13.21mm	U.S. Hall Rifle and Carbine, 1834-1842	.525in 13.34mm	219grs 14.19gm	75grs 4.86gm		2.1in 53.34mm	Specimen—Lewis
PBL 3	16.89mm .665in	Norwegian Rifle, Model 1842	17.68mm .696in	40gm 617.3grs	6gm 92.6grs		59.69mm 2.35in	Specimen
PBL 4	.56in 14.22mm	U.S. Merrill, Model 1859	.557in 14.15mm	400grs 25.92gm	40grs 2.59gm		1.65in 41.91mm	Specimen—Lewis
PBL 5	.63in 16mm	Unknown	.63in 16mm				1.90in 48.77mm	
PBL 6	.56in 14.22mm	U.S. Sharps, Model 1858	.575in 14.61mm	465grs 30.13gm	55grs 3.56gm		2.50in 63.50mm	Specimen—Lewis
PBL 7(a)	.52in 13.20mm	U.S. Sharps, ca. 1861	.540in 13.72mm	450grs 29.16gm	60grs 3.89gm		2.10in 53.34mm	Specimen—Lewis
PBL 7(b)	.52in 13.20mm	U.S. Sharps, ca. 1861	.540in 13.72mm	450grs 29.16gm	60grs 3.89gm		2.20in 55.88mm	Specimen
PBL 8	.52in 13.20mm	U.S. Sharps Carbine, ca. 1861	.540in 13.72mm				1.92in 48.77mm	Specimen
PBL 9	.54in 13.72mm	U.S. Starr Carbine, Model 1858	.555in 14.10mm	444grs 28.77gm	62grs 4.02gm		1.90in 48.26mm	Specimen—Lewis
PBL 10	.54in 13.72mm	U.S. Lindner Rifle, Model 1859 (altered)	.564in 14.33mm				3in 76.2mm	Specimen
PBL 11	.58in 14.73mm	U.S. Lindner Carbine, Model 1859	.587in 14.91mm				3.2in 81.28mm	Specimen

PERCUSSION BREECH-LOADING ARMS AND CARTRIDGE DATA

No.	Bore	Model and Date	Bullet Diameter	Bullet Weight	Powder Charge	Case Length	Cartridge Length	Sources
PBL 12	.58in 14.73mm	Unknown	.579in 14.71mm				2.35in 59.65mm	Specimen
PBL 13	12mm .472in	French Manceaux & Viellard, 1861	12.95mm .510in	30gm 463grs	5gm 77.2grs		76.50mm 3.02in	Ch. Ne, E. Timmermans, July 1967 Gun Report
PBL 14	.577in 14.65mm	English, Mont-Storm Carbine, 1861	.585in 14.86mm	325grs 21.1gm	69.4grs 4.5gm		2.13in 54.1mm	Specimen—Mattenheimer
PBL 15	.577in 14.65mm	English, Prince's Rifle, 1859	.600in 15.24mm				1.75in 44.45mm	Specimen
PBL 16	.56in 14.22mm	U.S. Colt Revolving Rifle, ca. 1862	.565in 14.35mm	490grs 31.75gm	45grs 2.92gm		1.75in 44.45mm	Specimen
PBL 17(a)	.56in 14.22mm	U.S. Colt Revolving Rifle, ca. 1862	.580in 14.73mm	480grs 31.1gm	55grs 3.56gm		1.75in 44.45mm	Specimen—Lewis
PBL 17(b)	.52in 13.21mm	U.S. Sharps Carbine, ca. 1862	.540in 13.72mm	450grs 29.16gm	50grs 3.24gm		1.80in 45.72mm	Specimen—Lewis
PBL 17(c)	.52in 13.21mm	U.S. Hall Carbine, ca. 1862	.54in 13.72mm	435grs 28.19gm	60grs 3.89gm		2.07in 52.58mm	Specimen—Lewis
PBL 18	.54in 13.72mm	U.S. Greene Rifle, 1857	.546in 13.87mm	546grs 35.38gm	68grs 4.41gm		2.5in 63.5mm	Lewis
PBL 19	13.2mm .520in	Russian "Obturator" Rifle, 1860	13.46mm .530in				49.53mm 1.95in	Mattenheimer, Schmidt
PBL 20	.577in 14.65mm	English Calisher & Terry Rifle, 1858	.575in 14.60mm				2.19in 55.63mm	Specimen
PBL 21	13.9mm .547in	Saxonian Cavalry Carbine, 1865	13.97mm .550in				48.26mm 1.90in	Specimen—Thierbach

PBL 22	.577in 14.65mm	English Wilson Rifle, 1860	.580in 14.73mm				2.48in 62.99mm	Specimen
PBL 23	.442in	English Westley Richards Carbine, 1862	.455in 11.56mm				2.60in 66.04mm	Specimen
PBL 24	.54in 13.72mm	U.S. Marston Rifle, 1853	.545in 13.84mm				1.40in 35.56mm	Specimen
PBL 25	.36in 9.14mm	U.S. Marston Rifle, 1853	.347in 8.81mm				1.68in 42.67mm	Specimen
PBL 26	14.5mm .570in	Bavarian Lindner Rifle, 1867	14.60mm .575in				60.45mm 2.38in	Specimen
PBL 27	14.5mm .570in	Bavarian Lindner Carbine, 1867	14.48mm .570in				53.34mm 2.10in	Specimen
PBL 28	14.5mm .570in	Bavarian Lindner Explosive, 1867	14.48mm .570in				60.45mm 2.38in	Specimen
PBL 29	.50in 12.70mm	U.S. Smith Carbine, 1857	.520in 13.21mm	368grs 23.85gm	50grs 3.24gm	1.54in 39.1mm	1.91in 48.51mm	Specimen—Lewis
PBL 30	.50in	U.S. Smith Carbine, 1863	.520in 13.21mm	376grs 24.36gm	40grs 2.59gm	1.34in 34.03mm	1.83in 46.48mm	Specimen—Lewis
PBL 31(a)	.54in 13.71mm	U.S. Burnside, Foster Impr., 1860	.565in 14.35mm	400grs 25.92gm	53grs 3.43gm	2.00in 50.80mm	2.55in 64.77mm	Specimen—Lewis
PBL 31(b)	.54in 13.71mm	U.S. Burnside, common form	.565in 14.35mm	400grs 25.92gm	53grs 3.43gm	1.89in 48mm	2.38in 60.45mm	Specimen—Lewis
PBL 32	.54in 13.71mm	U.S. Burnside Carbine, 1863	.560in 14.22mm	370grs 23.98gm	40grs 2.59gm	1.90in 48.26mm	2.3in 58.42mm	Specimen—Lewis

PERCUSSION BREECH-LOADING ARMS AND CARTRIDGE DATA

No.	Bore	Model and Date	Bullet Diameter	Bullet Weight	Powder Charge	Case Length	Cartridge Length	Sources
PBL 33	.50in	U.S. Maynard Carbine, 1861	.52in 13.21mm	350grs 22.68gm	43grs 2.79gm	1.21in 30.73mm	1.75in 44.45mm	Specimen–Lewis
PBL 34	.50in 12.70mm	U.S. Maynard Carbine, 1859	.52in 13.21mm	342grs 22.16gm	50grs 3.24gm	1.19in 30.22mm	1.87in 47.50mm	Specimen–Lewis
PBL 35	.50in 12.70mm	U.S. Gallager Carbine, ca. 1862	.52in 13.21mm	435grs 28.19mm	60grs 3.89gm	1.67grs 42.42gm	2.00in 50.8mm	Specimen–Lewis Specimen–Lewis
PBL 36	.50in 12.70mm	U.S. Gallager Carbine, 1860	.525in 13.34mm	374grs 24.23gm	50grs 3.24gm	1.69in 42.93mm	2.12in 53.8mm	Specimen–Lewis
PBL 37	.69in 17.53mm	U.S. Maynard Capsule, 1857	.68in 17.27mm		70grs 4.5gm	1.25in 31.75gm	1.56in 39.62mm	Specimen–Lewis
PBL 38	.69in 17.53mm	U.S. Maynard "Flop Ear," 1857	.68in 17.27mm		70grs 4.54gm	1.28in 32.51mm	1.81in 45.94mm	Specimen–Lewis
PBL 39	.69in 17.53mm	U.S. Maynard Cord Extractor, 1857	.68in 17.27mm		70grs 4.54gm	1.28in 32.51mm	1.81in 45.94mm	Specimen–Lewis
PBL 40	.54in 13.72mm	U.S. Hunt, 1848	.550in 13.97mm				1.05in 26.67mm	Specimen
PBL 41	.56in 14.22mm	U.S. Billinghurst, ca. 1861-1865	.545in 13.84mm	390grs 25.27gm	93grs 6.03gm	2.05in 52.07mm	2.55in 67.77mm	Specimen–Lewis
PBL 42	.58in 14.73mm	U.S. Ager (Union) Machinegun, 1861	.577in 14.66mm	465grs 30.13gm	40grs 2.59gm		2.75in 69.85mm	Specimen (the cylinder)
PBL 43	.56in 14.23mm	Unknown	.570in 14.48mm				2.56in 65.02mm	Specimen

Note: Diameters given are actual size of projectile as reported by sources cited and confirmed by measurement where possible.

8 THE TUBE PRIMED BREECH-LOADER

Cartridge development during the first half of the 19th century did not progress in neat, chronological order. The various systems reviewed here overlapped the separate-primed, breech-loading phase. Some appeared and fell into obscurity long before other less advanced ideas achieved a measure of acceptance.

One of those that never made it was the tube primed cartridge and rifle that used it. Joseph Manton, an English gunsmith, invented the thin copper tube containing detonating powder in 1818. This device was inserted into the priming vent, or touch-hole, with the outer half lying across an anvil surface against which the hammer nose crushed it. Some were designed to be crushed and exploded in the center. By 1831, the French inventor Augustus Demondion had incorporated this priming tube in the cartridge body itself for use in a breech-loading single-shot rifle. The French gunmaker Robert also developed a simi-breech-loader. Both were undoubtedly shown to French military authorities and the Robert was tested at Harpers Ferry Arsenal in the United States in 1837. However, the time for self-contained ammunition and breech-loaders for it had not arrived. Military leaders believed that incorporating primer with black powder in a cartridge was likely to cause accidental explosions.

Genhart tube ignition cartridge for pistol.

Jaroslav Lugs states that the Czech arms inventor, Sylvester Krnka, designed a similar cartridge in 1849 for one of his early breech-loading actions. In this one, the detonating tube was attached to the bullet by a wire. The bullet carried the priming tube with it out the muzzle at the shot. Thierbach discusses similar cartridges using tubular or spherical (?) percussion devices developed by Montigny (a Belgian) and Prince Alexander von Wurttemberg (Germany) in 1833 and 1835 respectively. These remain unidentified. Thirty years later, Dr. Edward Maynard, the American inventor, patented a tube-primed cartridge (No. 45,420 dated December 13, 1864) in which the tube projected from the side of the case near the head similar to a pinfire. The following year he patented a breech-loading action for it, but neither went beyond the experimental stage.

TPBL 1 – 16.50mm (.65in) French
Demondion, 1831

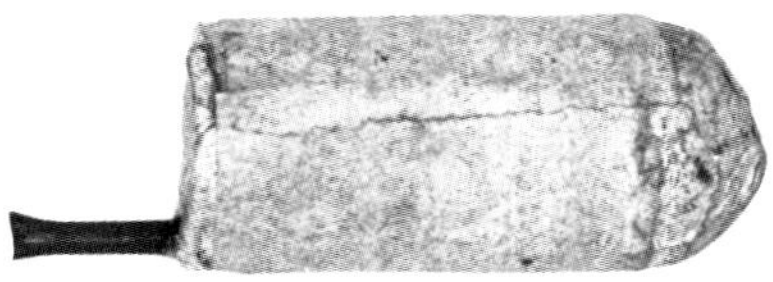

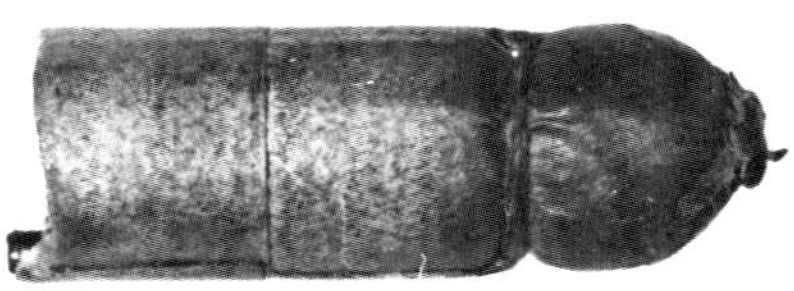

Description

(a) Grey paper tube pasted closed over a round ball. An extra wrap of paper covers the case body and a patch of the same paper is cemented over a cardboard disc to close the powder column. A copper tube filled with fulminate composition, pinched closed at its exterior end, protrudes from the powder column at the edge of the base.

(b) Same construction as above, paper discolored brown. The tube is protected in this specimen by a hollowed wooden plug covered with paper and wrapped to the cartridge body. This differs from (a) which has a removable slip-on paper covered wooden plug to protect the tube before use. The white band is an identification marker wrapped about the body by the present owner.

(c) A variant of the two above with brown paper case tied over a round ball with thread tie between powder and ball. The thread extends to hitch over the copper priming tube. The base is closed with a brown cardboard disc and wrapping of paper cemented on over the thread holding the priming tube in place.

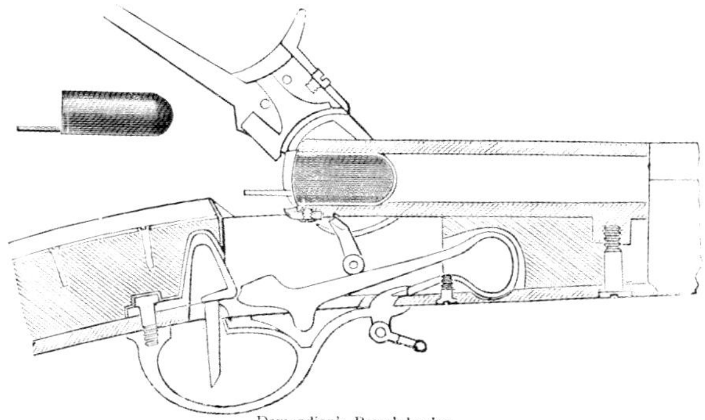

Demondion's Breech-loader.

The Demondion rifle, described and illustrated in English patent No. 6173 dated July 13, 1839, employed a hinged breechblock and underhammer. Lifting up on the operating lever attached to the breechblock lifted the block to expose the chamber and also depressed the mainspring hammer situated beneath the barrel until it engaged with the sear. When the trigger was pulled, the sear released the hammer which flew up to strike the priming tube in the cartridge against the base of the breechblock. This exploded the fulminate composition in the tube which was transmitted to the powder charge.

The Robert rifle, using the same type cartridge, employed a trigger guard underlever which allowed the breechblock to drop down, exposing the chamber for loading. (Drawing from W. W. Greener's "The Gun and Its Development," 7th ed., 1899)

References: "The Gun and Its Development, W. W. Greener, 7th ed., 1899
"Sharps Firearms," Frank Sellers, 1978
"Firearms Past and Present," Jaroslav Lugs, 1973
"Die Geschichtliche Entwicklung der Handfeuerwaffen," Moritz Thierbach, 1886-1899
"The International Cartridge Collector," John Belton, editor, September/October 1978

TUBE PRIMED BREECH-LOADING ARMS AND CARTRIDGE DATA

No.	Bore	Model and Date	Bullet Diameter	Bullet Weight	Powder Charge	Cartridge Length	Sources
TPBL 1(a)	16.50mm .65in	Demondion for Roberts Rifle, 1831	17.15mm .675in			50.04mm 1.97in	Specimen
1(b)	16.50mm .65in	Presumed for same rifle	16.31mm .642in			72.50mm 2.85in	Specimen
1(c)	16.50mm .65in	Presumed same rifle (?)	17.07mm .672in			50.80mm 2.00in	Specimen

Note: Bullet dimensions taken over paper case.
Actual diameter of projectile would be .10 to .20 thousandths less depending on thickness of case material.

9 THE NEEDLEFIRE BREECH-LOADER

Johann Nikolaus von Dreyse, a German engineer, is credited with developing the first practical military needlefire cartridge and a revolutionary bolt action which formed the basis for all of the metallic cartridge bolt action rifles that appeared during the latter half of the 19th century.

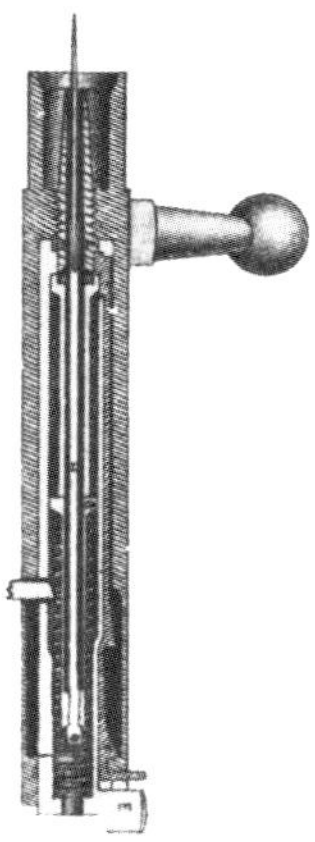

Needle rifle bolt

In the 1820's after working briefly with Samuel J. Pauly in France, von Dreyse commenced experiments in his home city of Sommerda, evidently inspired by Pauly's 1812 application of the percussion system in producing the first metallic self-contained centerfire cartridge. Initially von Dreyse concentrated his efforts on perfecting a muzzle-loading ball containing a percussion primer. The weapon he designed to use it with was fitted with a breech assembly containing a spring loaded needle which was cocked by means of a lever on the right side of the breech. When the trigger was pulled, the sear released the needle which drove forward under the sping pressure to penetrate the powder charge and strike the primer seated on the base of the ball. He submitted this idea to the War Ministry in 1827 but it was rejected. The system was prone to accidental discharge during loading if the ball was rammed down too hard.

The Self-Contained Needlefire Cartridge

After this initial failure, von Dreyse turned his efforts toward perfecting a breech-loading action and a cartridge containing its own priming at the base of the projectile. In 1833 he submitted the first of his models to the War Ministry and continued to perfect it until it was ultimately adopted in December 1840. This rifle was designated the Model 1841 and von Dreyse, who received the contract to manufacture it, delivered the first of them in late 1842.

The cartridge von Dreyse developed for this rifle was loaded with a 15.2mm round ball seated in a cylindrical papier-mache sabot hollowed out in front and containing a priming pellet of mercuric fulminate in a cavity in its base. This ball was near bore size. However, the military soon changed this to an oval ball 13.6mm in diameter so that the papier-mache sabot was the only part of the projectile which touched the surface of the bore.

NFR 1 – 15.43mm (.607in) Prussian
Needle Rifle, Model 1847-1855

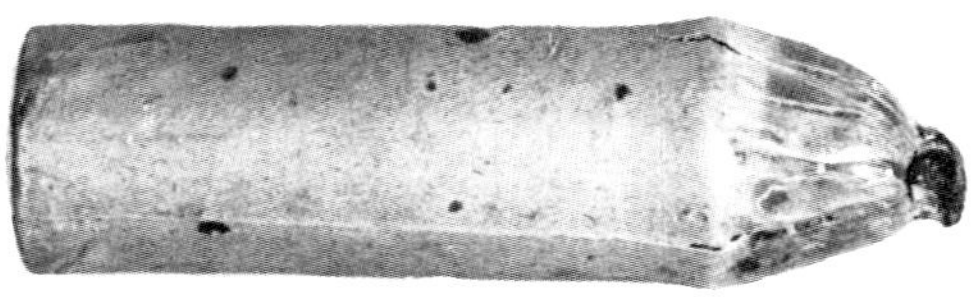

Description

Cardboard tube covered with a wrap of white paper tied closed over the nose of an oviform (egg-shaped) bullet which sits with its rear half in a bore-size papier-mache cylinder hollowed out to a depth of two-thirds its length to the shape of the bullet. A disc-shaped pellet of fulminate of mercury occupies a central cavity in the base of the papier-mache "sabot." Black powder fills the rear one half of the case which is closed with a cardboard disc partly covered by the folded and cemented rear edges of the case wrapping. Bullet diameter is actually 13.6mm (.535in). (Drawing by M. E. Whiteley)

NFR 2 – 15.43mm (.607in) Prussian
Needle Rifle, Model 1872

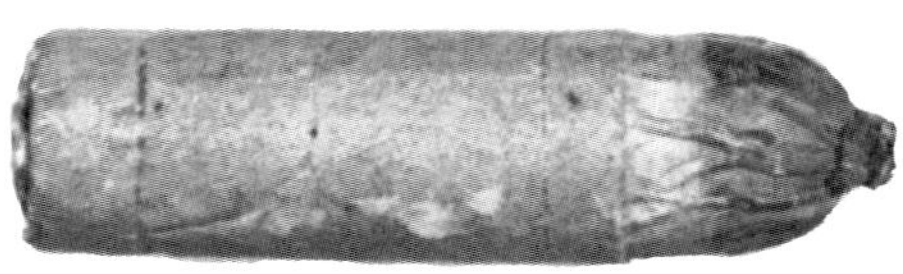

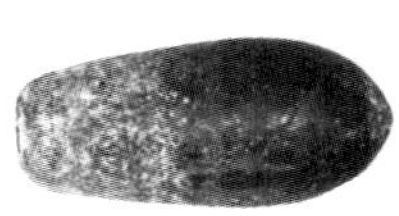

Same construction and appearance as above except 5mm shorter overall. The bullet in the M72 cartridge is 12mm (.472in) in diameter. This "Beck" improvement was adopted after the Franco-Prussian War (1870-1872) to improve the muzzle velocity and range of the needle rifle.

Comments

The first Dreyse needle-fire rifle cartridges used in the original Model 1841 adopted by Prussia were loaded with round ball. Commencing with the 1847 model, the egg-shaped bullet was retained throughout the life of this rifle, although bullet shape was altered from a more pointed style used in the 1847 to the round-nosed type shown above adopted in 1855. Since the bullet is covered, it is difficult to tell these cartridges apart. Specimens are found with dark blue case. Some were made with explosive and incendiary ball.

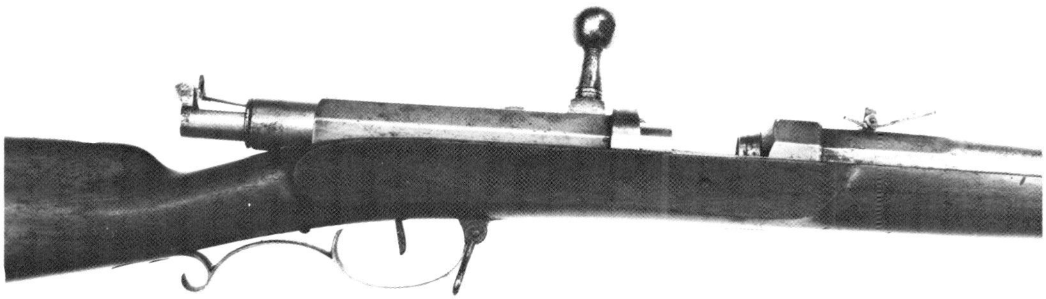

The illustrations of von Dreyse's needle rifle action shows that the cone-shaped breech end of the barrel fit into the face of the bolt when the rifle was closed. This directed any gas leakage from the exploding powder charge toward the front away from the shooter's face. Repeated firing of the rifle evidently eroded these close-fitting parts, and it has been frequently written that leakage became so bad in some of these guns that the soldiers were afraid to fire them from the shoulder. However, Walter reports that a recent test of the Model 1841, the first pattern of this rifle, showed gas leakage to be insignificant. Prussian authorities accepted several hundred thousand of these arms through some ten models over a 35-year period.

The long, thin needle which frequently broke from passage through the charge and the eroding effect of exploding powder, was a weakness in this rifle and the soldier carried replacements. The combustible paper case also left residue in the chamber and the first models were made with a hollow bolt face so that residue would blow back into this chamber and be carried out when the bolt was drawn back. Later, the Beck improvement shown in the lower illustration here used a solid-faced bolt fitted with a leather washer to effect obturation. Unburned case material must have then been pushed out ahead of the next cartridge loaded. The Beck model is identifiable from the screw head on top of the bolt. (Milwaukee Public Museum photo)

References: "Firearms Past and Present," Jaroslav Lugs, 1973
"The German Rifle," John Walter, 1979
"Das Zündnadel-Gewehr," W. von Plönnies, 1865

NFR 3 – 11mm (.433in) French Chassepot, 1866

Description

(a) Tan paper tube enclosing a round-nose, elongated, paper-patched bullet, a felt washer at the bullet's base, followed by a charge of black powder. At the rear of the powder charge is seated a percussion primer with its interior surface facing to the rear. The primer is held in the center of a papier-mache washer and its base is covered by a thin gutta percha disc 8mm in diameter. The paper case is folded over and cemented to the disc and the bottom of the washer.

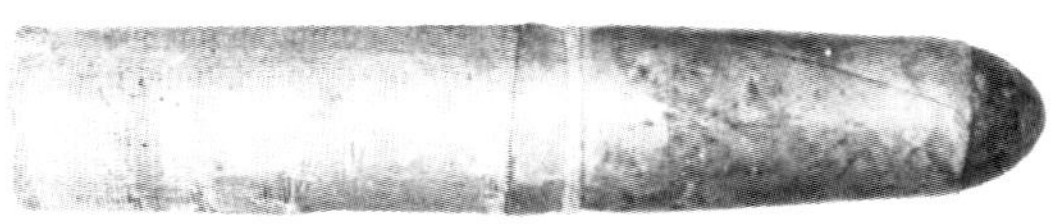

(b) Same general appearance as above except two-piece case. The tan paper bullet wrap is tied to a separate powder container, also of paper but covered with fine mesh linen.

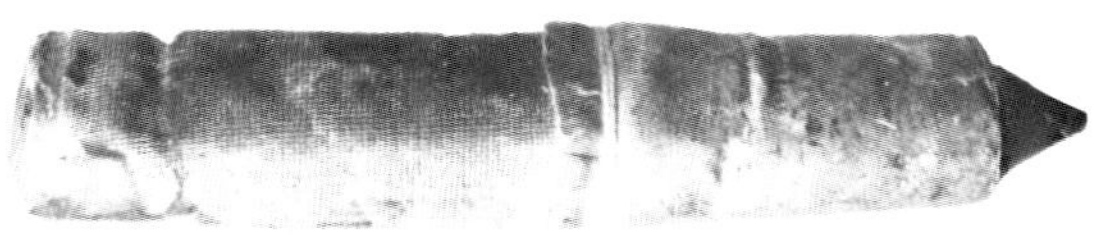

(c) Same construction as above, loaded with a steel-tipped lead bullet.

Comments

The French Chassepot cartridge, with its base-situated primer, obviated the necessity for the firing needle to penetrate the powder charge, although it still had to penetrate the base of the case about one-fourth inch to reach and explode the primer. The bullet, made with a narrow base band of lead somewhat larger than the bore, received its stabilizing spin directly from the rifling.

Although France was some 25 years behind Prussia in adopting the needlefire, breech-loading system, it was the revolutionary small bore and comparatively long projectile for its weight used in the Chassepot which led the Prussians to modify their own cartridge after the 1870-1872 Franco-Prussian War.

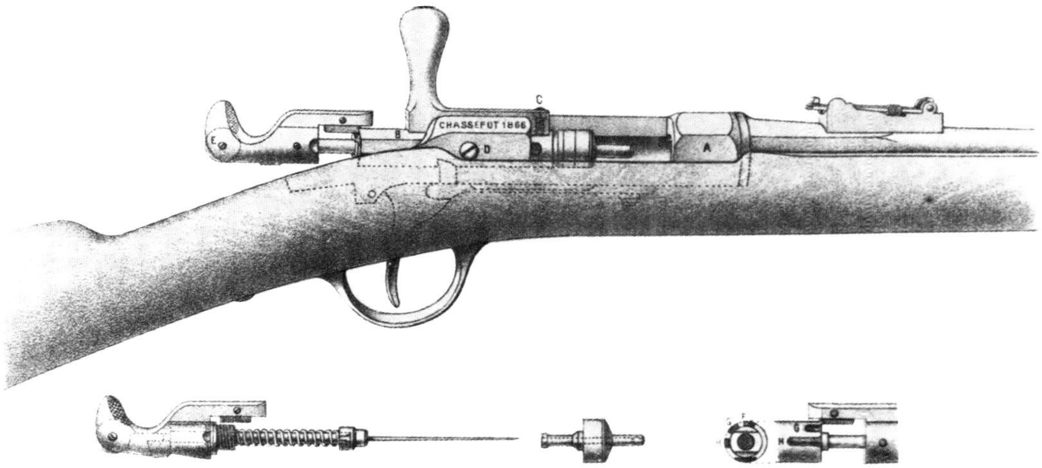

Antoine Alphonse Chassepot, employed at a French artillery arsenal, designed this needlefire rifle which was adopted by France in 1866. The Fusil Mod. 66 incorporated the basic features of the von Dreyse needle rifle and improved on it to some extent. The needle, although still long and relatively fragile, did not have to travel through the powder charge as in the von Dreyse arm because the primer was located near the base of the cartridge. This eliminated the problem of needle erosion and breakage for the most part. Chassepot also fitted a rubber obturating disc on the head of the bolt. This piece tended to erode from the effects of exploding powder and allowed gas to leak out through the action. Therefore, the soldier carried replacements.

British military authorities tested the Chassepot, among other arms, during the 1860's when they were seeking a breech-loading replacement for the Enfield rifle. The resulting report included the observation that repeated firings of the Chassepot so clogged the chamber with powder residue and unburned cartridge case material that it became necessary to pound the cartridge into the chamber enough to close the bolt. (Drawing from Rodolphe Schmidt's "Les Armes A Feu Portatives," 1877)

Reference: "Military Breech-loading Rifles," V. D. Majendie and C. O. Browne, 1870 (reprinted 1973 by Arms and Armour Press, London with Fortress Publications, Inc., Stoney Creek, Ontario, Canada)

NFR 4 – 17.5mm (.689in) Italian Carcano, 1867

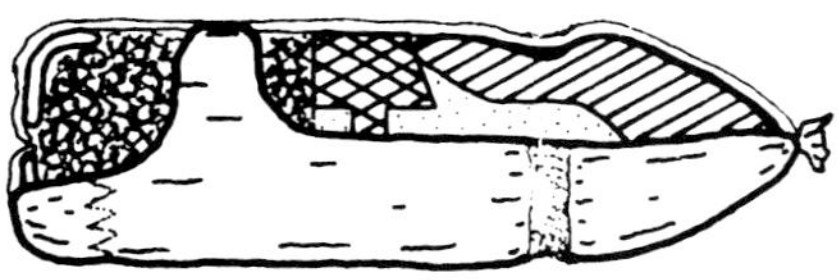

Description

Beige paper tube tied over the nose of a hollow-base, single-groove conical bullet seated on a papier-mache sabot which holds the primer in a central base cavity. Remainder of case filled with black powder and closed with a cardboard base disc centrally perforated. Another wrapping of case paper is cemented around the lower half of the cartridge and an additional patch with fingers around the lower case body covers the perforated base disc.

Comments

The Carcano cartridge followed the Dreyse system only in having the primer at the front of the powder charge. Otherwise, it employed an expansion bullet which engaged the rifling, the sabot serving only to firmly hold the priming pellet in place. (Drawing by Richard A. King)

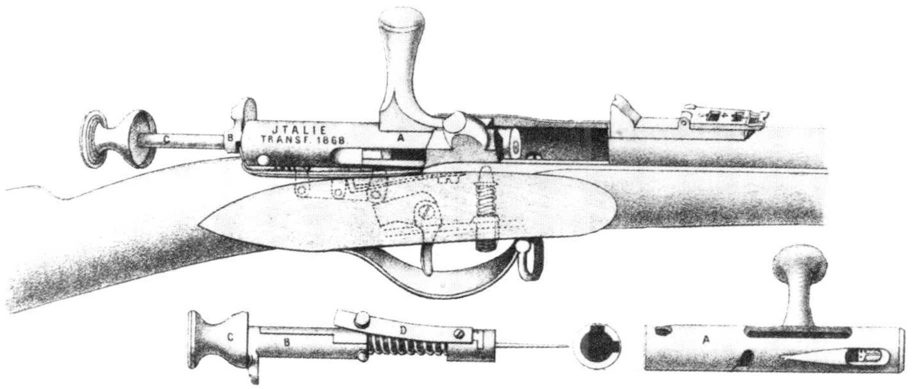

The Carcano needle rifle, named after the Italian inspection engineer at Turin arsenal who designed it, was adopted by that country in 1867 for the purpose of converting existing muzzle-loading arms to breech-loaders. The chamber section of the muzzle-loading barrel was cut off and this bolt action system fitted to it. The Carcano had the same drawback as the von Dreyse—a long needle that had to penetrate the powder charge to strike the primer. It had no greater range than the muzzle-loader and was soon replaced by rapid advancements in firearm design when Italy adopted a 10.4mm centerfire rifle in 1871. (Drawing from Schmidt's "Les Armes A Feu Portatives," 1877)

NFR 5 – 15.24mm (.600in) Russian
Carle (also Karl), 1867

Description

Cardboard case tube enclosing just over one-half the length of an elongated, three-groove, hollow-base bullet containing an iron expansion plug, or "culot." The case material is pressed into the bullet's grooves and cannelured below its base. The powder charge is in contact with the bullet. Four elements comprise the base of the cartridge: a percussion primer situated open face to the base; a disc washer of cardboard fitted over the primer and cemented to a thicker perforated base disc of cardboard larger in diameter than the case body which forms a rim. The whole is held together at the base by several wrappings of thread around the base above the rim, and the case body material is folded over and cemented to the base disc, leaving a central opening through which the needle may pass to reach the primer. (Drawing by M. E. Whiteley)

Comments

The Russian Carle cartridge, designed by Colonel Weltischtschew was a much more sophisticated product than the Prussian, French, or Italian types. The base construction and location of the primer resulted in a shorter travel for the needle, rigidly held the primer in place, and provided for breech obturation since the combination of cardboard discs surrounding the wad, plus the rim, were compressed against the bolt face and chamber wall at the moment of powder ignition. However, it still was not a foolproof breech obturator because of its cardboard case construction. The case material was expected to be all or partly consumed on firing.

Reference: "Infanterie-Präzisionswaffen," Hermann Weygand, 1872

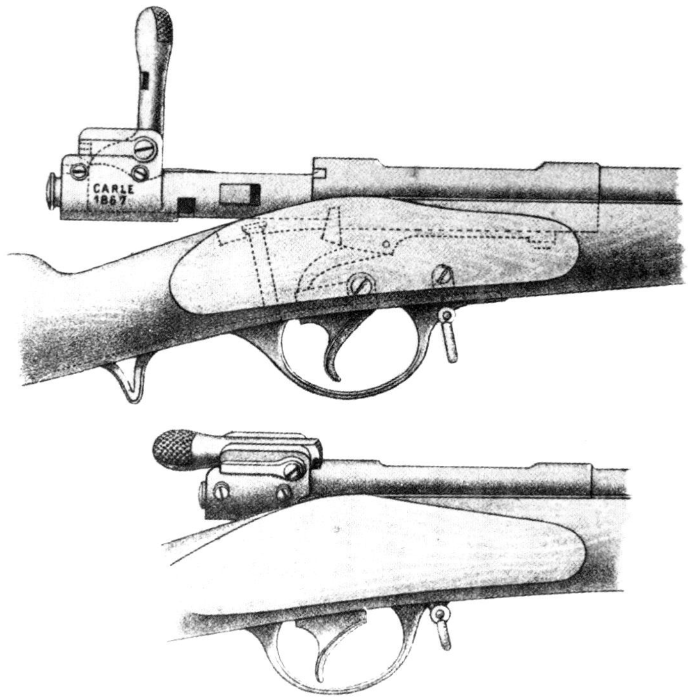

A Suhl gunmaker named Carle designed this action which was adopted by Russia in 1867 to convert its existing rifled muzzle-loaders to breech-loaders. A bolt action, it differed from the von Dreyse needle rifle. The operating handle on the bolt head was pushed forward to unlock the action and this allowed the bolt, which did not turn, to be moved to the rear for loading. Pushing it forward and laying the operating lever down to the rear locked the action for firing. The Carle also used a shorter firing needle to reach the primer seated in the base of the cartridge.

The Russians evidently were applying every new idea they saw to convert muzzle-loading rifles at that time. About the time this rifle was adopted, they began converting muzzle-loaders with the Berdan breech-loading system and in 1869 the Krnka breech-loader was also adopted. Both used metallic cartridges. However, 200,000 of these Carle needlefire guns were issued in 1872 to the army of the Caucasus and to various military districts. (Drawing from Rodolphe Schmidt's "Les Armes A Feu Portatives," 1877)

Reference: Lugs' "Firearms Past and Present," English edition, 1973

NFR 6 – 12.2mm (.480in) Swedish Haagstrom (also Hagstrom), Model 1867

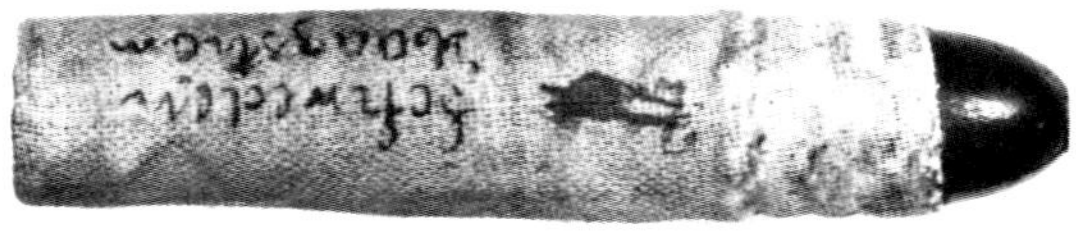

Description

Linen case enclosing a conical flat-nosed, three-groove bullet with hollow base and powder charge filled in directly over the bullet. A metal culot, or cup, supporting a percussion primer held in by papier-mache wad occupies the base of the cartridge. The culot has a central perforation to admit the flame from the exploding primer in to the powder. The handwritten label on the case reads "Schweden Haagstrom."

NFR 7 – 10.4mm (.409in) Unknown

Brown paper case containing a round-nosed, paper-patched bullet secured in the case mouth by a wrapping of string. Attached to the base is a cardboard disc larger in diameter than the case body which may have served as a rim. The disc is centrally perforated and a primer appears to be situated within the cartridge near the base (partly visible through the damaged case wall).

Comments

The Swedish Haagstrom cartridge closely resembled the French Chassepot, but the primer is more firmly held in the base by the metal culot and papier-mache wad. The needle had only to penetrate about one quarter inch to strike the primer.

The second specimen shown above may be a Swiss experimental since it is the same caliber as the Federal rifle paper cartridge and the later Vetterli rimfire. However, this must await confirmation.

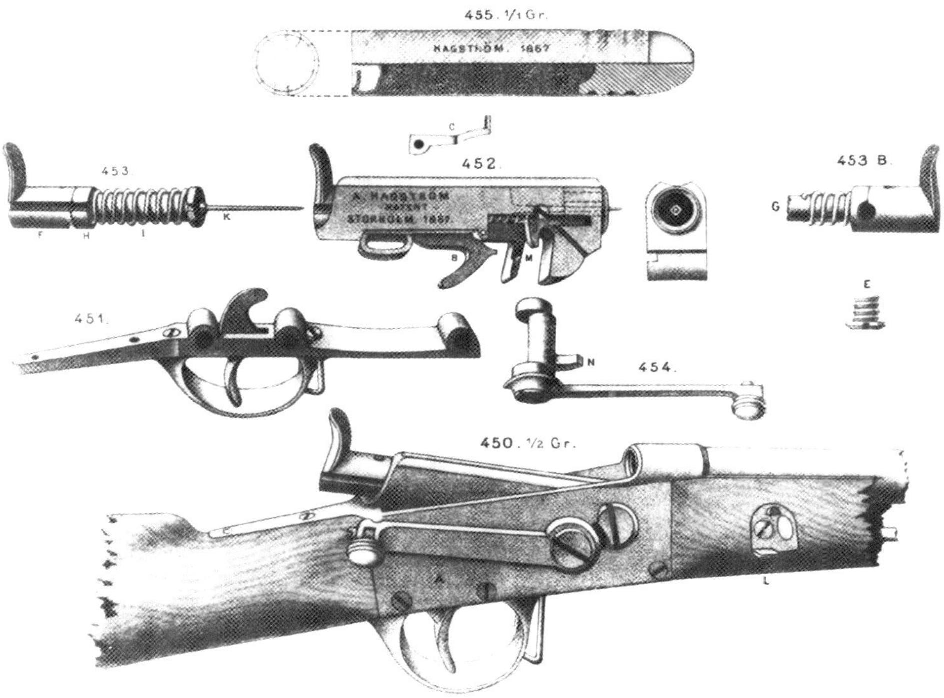

The Haagstrom rifle used for a few years by Sweden was operated by a lever on the right side of the frame which was attached to a bolt running through the action from right side to left. This bolt was made with a lifter arm which operated against the bottom of the breechblock. To open the action, the soldier drew back the bolt containing the firing needle by the knurled cocking piece. The bolt was engaged by the sear. Then the operating lever was rotated counterclockwise to the rear. This allowed the breechblock to drop down at the front, exposing the chamber for loading. After the cartridge was inserted, the lever was moved clockwise to the forward position. This brought the breechblock back up in line with the chamber. A hard rubber obturating disc appears to have been fitted to the rifle's bolt face. The metal culot in the cartridge base filled with the primer and a papier-mache wad also was expected to expand under the force of the exploding powder gases and seal the breech. (Drawing from Rodolphe Schmidt's "Les Armes A Feu Portatives," 1877)

NEEDLEFIRE BREECH-LOADING ARMS AND CARTRIDGE DATA

No.	Bore	Model and Date	Bullet Diameter	Bullet Weight	Powder Charge	Cartridge Length	Sources
NFR 1	15.24mm .600in	Prussian Needle Rifle Models 1847-1855	13.6mm .535in	31gm 478grs	4.9gm 75.6grs	62-64mm 2.4-2.5in	Specimens
NFR 2	15.24mm .600in	Prussian Needle Rifle Model 1872 (Beck)	12mm .472in	21gm 324grs	4.5gm 69.4grs	57-59mm 2.2-2.3in	Specimen—von Plonnies
NFR 3(a)	11mm .433in	French Chassepot, 1866	11.6mm* .457in	25gm 386grs	5.6gm 86.4grs	68.58mm 2.70in	Specimen—Weygand
NFR 3(b)	11mm .433in	Same as above	11.6mm* .457in	”	”	72mm 2.83in	Specimen
NFR 3(c)	11mm .433in	Same as above	11.6mm* .457in	”	”	71.5mm 2.81in	Specimen
NFR 4	17.5mm .689in	Italian Carcano, 1867 Transformed Musket	17.47mm .688in	36gm 556grs	4.5gm 69.4grs	52.07mm 2.05in	Specimen—Schmidt
NFR 5	15.24mm .600in	Russian Carle, 1867	15.2mm .598in	35.5gm 548grs	5.07gm 78.2grs	62mm 2.44in	Weygand
NFR 6	12.2mm .480in	Swedish Haagstrom, 1867	13.18mm** .519in			70mm 2.75in	Specimen—Schmidt
NFR 7	10.4mm .409in	Unknown	10.41mm** .410			61.5mm 2.42in	Specimen

Notes: *Diameter over paper—belt on base of bullet is 11.86mm (.467in)
**Taken over paper covering largest diameter of bullet.

10 THE PINFIRE BREECH-LOADER

The original Lefaucheaux pinfire, 1835

Perfection of the percussion cap led to the idea of incorporating the cap within the body of the cartridge itself. H. Lefaucheaux of France patented a breech-loading drop barrel action for shotgun in 1835, along with a cartridge for it. His cartridge was a cardboard tube with metal head. A sturdy pin was inserted through a hole in the side of the head with its point set in a percussion cap seated against the interior opposite side of the head. The pin was held in place by a loop of fine wire through two holes in the base of the cartridge. When the Lefaucheaux gun was loaded, the cartridge pin protruded from the top of the joint between breech face and barrel through a small mortise cut in the mouth of the chamber. The hammer nose struck straight down on the pin and fired the internal cap, setting off the powder charge.

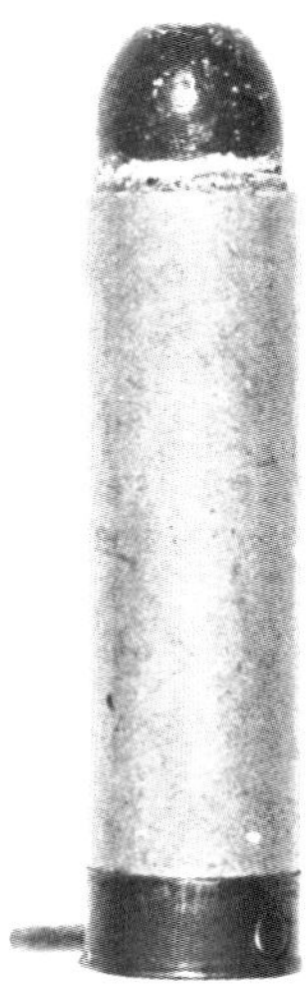

Chaleyer pinfire sporting rifle cartridge in .520in (13.2mm)

Houllier, another French inventor, improved on the Lefaucheaux cartridge with his 1846 patent for a compressed paper wad to be seated in the base of the head. This wad minimized gas leakage through the pin hole and the cavity for the percussion cap was placed in the center of the wad which also served to snugly hold the pin in place, thus obviating the need for a retaining wire. Chaleyer in his French patent No. 23,421 sacrificed this Houllier improvement by boring through both sides of the metal head. The cap sat flush with the outside wall of the head and the chamber served as the "anvil" against which the pin fired the priming cap. The idea was to make it easy to push out the spent cap with the pin and insert a new one. However, few of these were made.

The Houllier-improved pinfire was popular in Europe for use in shotguns, sporting rifles, and handguns up to the recent past in sizes from the diminutive 2mm for charm pistols up to at least 4 gauge in shotgun.

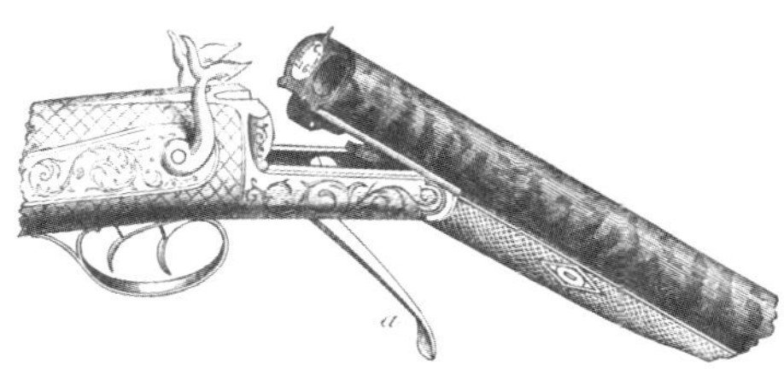

Lefaucheaux's gun

PFR 1 – 9mm (.354in) "Mousqueton des Cent Gardes," (Cavalry Rifle of the Hundred Guards), France, 1856

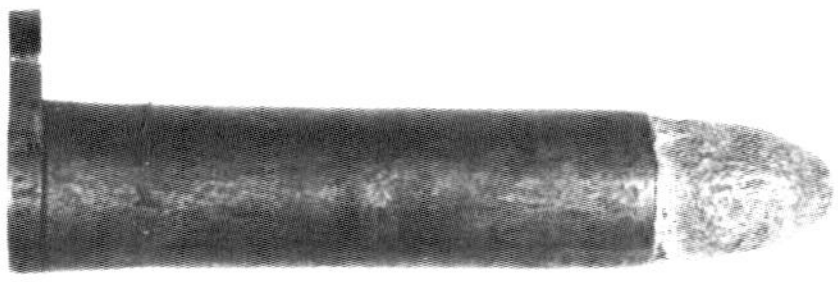

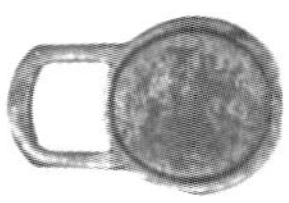

Description

Drawn brass tube closed at one end, with a brass ring and projecting stirrup pressed on over a reduced portion of the base. A U-shaped papier-mache wad situated inside the base holds the cap and detonating pin so that the base of the pin is at the bottom of the cartridge as illustrated here. A 30-grain powder charge and 190-grain grooved lead bullet completed the load.

Comments

The 1852 model of this cartridge was of the original Lefaucheaux pattern with cardboard tube case and metal head with exposed pin and similar positioning/extracting stirrup. The soldier chambered the cartridge so that the stirrup fitted over a protruding stud at the top of the receiver. He could also use this convenient handle as a means of extracting the fired case.

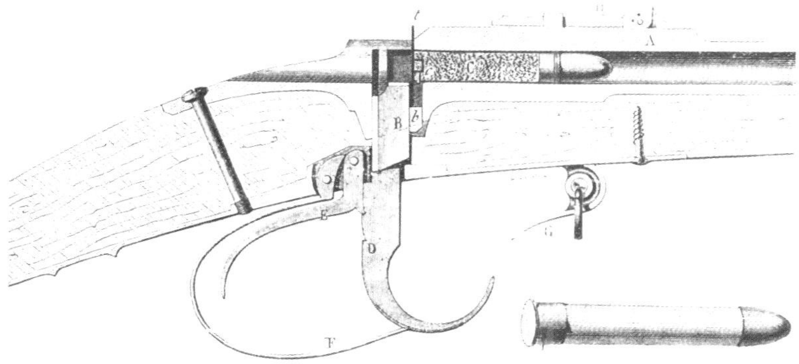

The Cent Garde cavalry rifle (also "Treuille de Beaulieu") had a vertical sliding block underhammer action. To operate, the soldier pulled down on the half-moon lever until it engaged the sear on the trigger. This held the action open and compressed the looping exterior spring curving around the trigger, its point against the bottom of the opening lever. The cartridge was then inserted stirrup up positioned over the stud at the top of the receiver. This placed the pin at 6 o'clock. When the soldier took aim, the action was still open, the breechblock having dropped down to admit the cartridge. When the trigger was pulled the sear disengaged from the notch in the underlever which was then forced upward under spring pressure, shoving the breechblock over the base of the cartridge. A stud at the front of the breeckblock struck the bottom of the cartridge driving the pin into the interior cap, setting off the charge.

This rifle was the smallest caliber military firearm of its day. Evidently the French determined that palace duty, much of it indoors, mandated an arm which would not do unwanted damage with a heavy, ricochetting bullet if the need arose for a shot to be discharged in the vicinity of occupied buildings. It was never used for field military purposes. (Drawing from W.W. Greener's "The Gun & Its Development," 7th ed., 1899)

PFR 2 – .58in (14.73mm) U.S. Gallager & Gladding, 1859

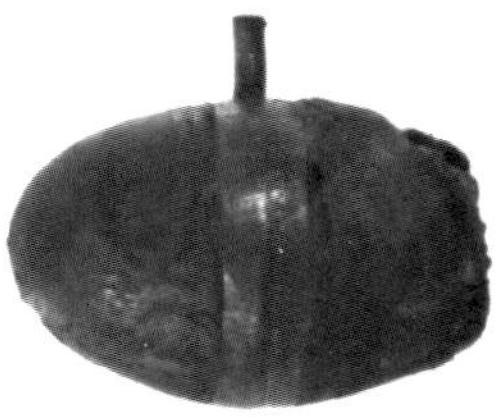

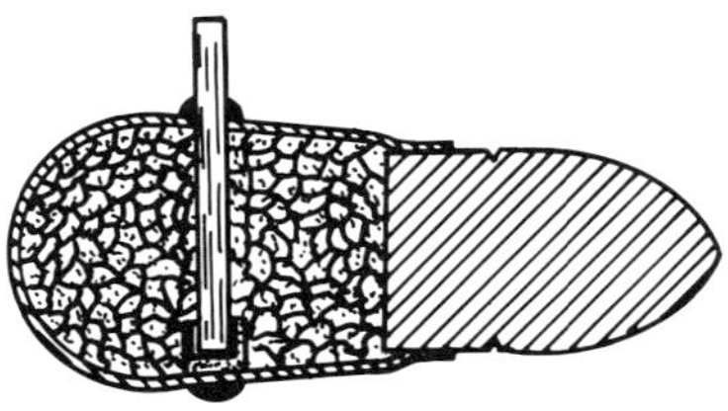

Description

The patent described this cartridge as ". . .case of paper or wood, or metallic shell, or wood or paper banded with metal. Pinfire . . ." The pin projected from the side of the bulbous case, in this specimen made of papier mache banded with metal. The other end of the pin was seated within a percussion cap and held it against the opposite interior wall of the case. A conical lead bullet loaded atop the powder charge filled in around the pin completed the loaded cartridge. (Drawing by M. E. Whiteley)

Comments

The Gallager & Gladding cartridge associated with the original patent and breech-loading carbine illustrated opposite is of different construction than the Schubarth cartridge discussed next. The difference lies in the projecting pin on the Gallager & Gladding as well as the case construction.

The empty case illustrated is the one found in the Gallager & Gladding carbine now in the Smithsonian Institution collection. (Smithsonian Institution photos, Patent Office collection)

The chamber was half in the pivoting block and half in the rear of the barrel in this carbine. The forward arm of the trigger guard served to latch the barrel assembly to the face of the breech when closed. Turning the trigger guard to the right moved this arm out of its notch at the bottom rear of the barrel formed with a hook, and allowed the barrel to drop of its own weight on the pin which held barrel and breech together, as in the common break-open shotgun. When the barrel dropped, the hinged breechblock tipped by its own weight to the rear, presumably bringing with it the empty fired case which could then be tipped or lifted out and a loaded cartridge inserted there. A mortise at the top of the barrel and breechblock allowed the cartridge's pin to protrude so it could be struck by the hammer.

None of the Gallager & Gladding carbines are known to have been used for military purposes. It is doubtful that more than trial specimens were made. (Smithsonian Institution photos, Patent Office collection)

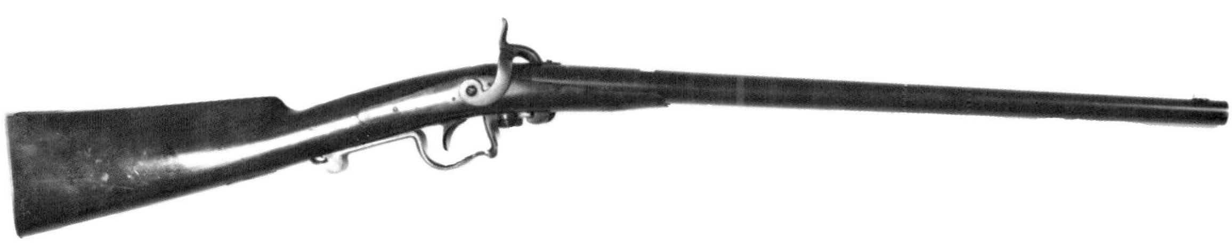

PFR 3 – .58in (14.73mm) Schubarth Rifle
U.S., 1859

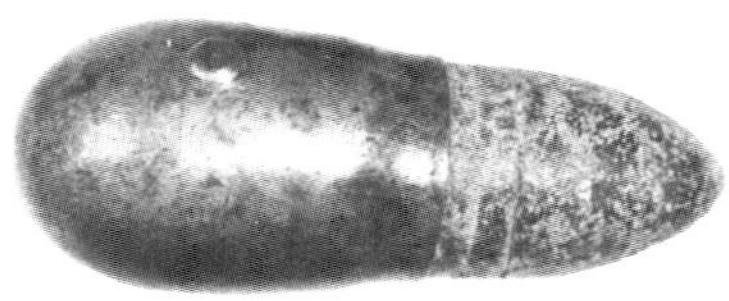

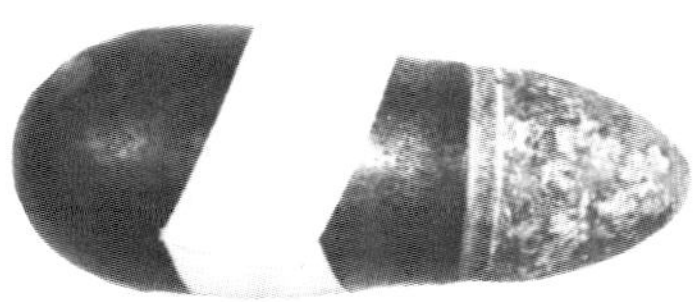

Description

(a) Drawn brass bulb-shaped case with a raised blister on one side. The detonating pin (visible in specimen shown with bullet removed) is fitted crosswise in the case with its head in the cavity formed by the blister and its point in a percussion cap held under it against the opposite wall. Black powder was filled in around the pin and cap, a wad seated on top, and a hollow-base Minie type grooved, lubricated bullet seated in the neck.

(b) Same case as above but loaded with a shorter bullet showing only one exposed groove. The white band is an identification strip placed on the case by the current owner of the specimen.

Comments

The Schubarth inside primed pinfire cartridge is commonly referred to as the Gallager & Gladding. However, the two are different. The Gallager & Gladding was made with the pin protruding from the side of the case and was described by the inventor as constructed with a case of paper or wood, or wood banded with metal, or as a metallic shell. The Schubarth cartridge is obviously an improvement of the original Gallager & Gladding patent. By placing the pin entirely within the case, a perfectly gastight, moisture proof case was achieved. The pin was just long enough to be held friction tight with its head in the blister on the case and its point in the percussion cap. The powder charge filled in around it kept it from being jarred out of place.

The Schubarth breech-loader was made after the Gallager & Gladding pattern with the same division of the chamber between the breechblock and the rear extremity of the barrel. To function with the bulbous cartridge having no projecting pin, a spring-loaded plunger on top of the receiver served as the striker against which the hammer dropped when the trigger was pulled. The cartridge had to be positioned with the blister directly beneath this plunger. (Smithsonian Institution photos, C.D. Schubarth Collection)

PFR 4 — 12mm (.472in) Lenoir Horizontal Pinfire, France, 1849

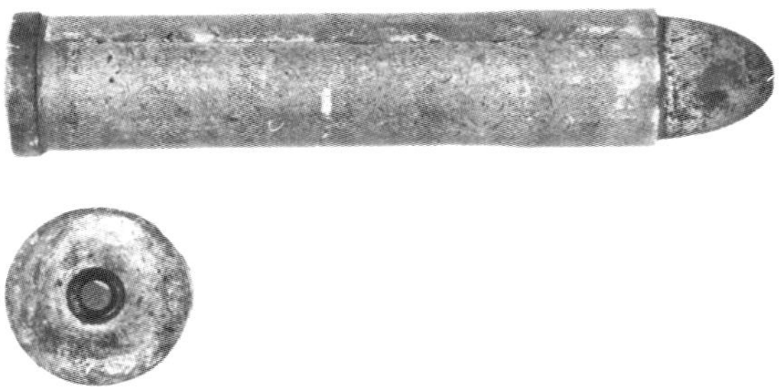

Description

Cardboard tube case with papier mache disc cemented to the base and forming a rim. A horizontal pin centered through a hole in the base disc extends through the powder charge to the percussion cap seated against the base of the bullet.

PFR 5 — 11.5mm (.453in) Pidault & Lagieze, France, 1867

Cardboard tube case with a brass base centrally perforated exposing a horizontal pin with point seated in a primer held by an internal cup embedded in a papier mache wad pressed down into the base of the case. Grooved lead bullet evidently with paper patch to the case mouth separating it from the powder charge. (Drawing by Richard A. King)

Comments

The Lenoir cartridge is one of a number of such experiments substituting the long needle of the needle rifle for a pin contained within the case body itself. Mattenheimer illustrates a specimen with the point of the pin in a cap at the base of the cartridge and the head of the pin wedged against the base of the bullet. None of these attained acceptance.

The Pidault & Lagieze cartridge was patented September 10, 1867 in the United States (No. 68,786), as well as in France where it was used in a bolt action, lever-operated rifle of experimental design with exposed hammer. Gevelot, the French ammunition maker, made this type cartridge with a brass cap over the entire base concealing the horizontal pin, and a drawn brass case variation is also known.

References: "Die Patronen der Ruckladungsgewehre," A. Mattenheimer, 1868 (reprint by Intersico Press AG, Zurich, Switzerland, 1975)
"Digest of Patents," Stockbridge, V.D., 1874

PINFIRE BREECH-LOADING ARMS AND CARTRIDGE DATA

No.	Bore	Model and Date	Bullet Diameter	Bullet Weight	Powder Charge	Case Length	Total Length	Sources
PFR 1	9mm .354in	Mousqueton des Cent Gardes, 1856	9.95mm .392in	12.31gm 190grs	1.94gm 30grs	43mm 1.69in	54.5mm 2.15in	Specimen
PRF 2	.58in 14.73mm	U.S. Gallager & Gladding, 1859	.60(?)in 15.24mm					
PFR 3	.58in 14.73mm	U.S. Schubarth, 1861	.61in 15.49mm	551grs 35.7gm	70grs 4.5gm	1.13in 33mm	1.90in 48.26mm	Specimen-Lewis
PFR 3(b)	.58in 14.73mm	Same as above	.61in 15.49mm	?	?	1.13in 33mm	1.77in 44.96mm	Specimen
PFR 4	12mm .472in	French Lenoir, 1849	12mm (?) .472in			61mm 2.40in	71.5mm 2.81in	Specimen
PFR 5	11.5mm .453in	French Gevelot, 1865–	11.55mm .454in	20.6gm 317.9grs	3.65gm 56.33grs	52.4mm 2.06in	65mm 2.60in	Gevelot Document

11 THE RIMFIRE CARTRIDGE

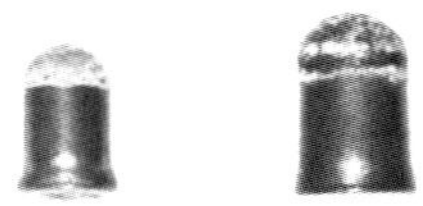

The 5 and 7mm Flobert salon pistol cartridges, also made in 4, 6, 8 and 9mm. A tiny 2mm rimfire was made for charm pistols.

The French gunsmith, Flobert, adapted the ordinary copper percussion cap to make a case for the self-contained cartridge he used in his 1849-patented breech-loading salon pistol. Dispensing with the right-angle "skirts" on the mouth of the percussion cap, he made a shell of thin copper spread out slightly at the base so that it would wedge just far enough into the pistol's chamber. The interior base was primed with fulminate mixture and the mouth was loaded with a round ball. This cartridge had only enough power for short-range indoor target shooting.

Horace Smith, an American, is known to have manufactured Flobert-type salon pistols in the United States during 1850 and 1851 and, therefore, was acquainted with the Flobert bulleted breech cap used in this arm. After an ill-fated effort to produce and promote an improvement on the Hunt and Jennings tubular magazine repeating rifle (See Percussion Breech-loader 40, p. 91), Smith and Daniel Wesson produced a .22 caliber (5.6mm) seven-shot revolver chambered for the Flobert cartridge. However, this little pistol was suitable only for indoor target shooting. So they lengthened the case a bit and formed it to produce a distinct rim, or flange, at the head. The drawing procedure left a small recess around the circumference of the rim between its inner surfaces. The case would now hold 2.5 grains of black powder. At first they tried holding the fulminate against the base of the cartridge case between thin discs but this did not work well. Finally, Wesson designed a tool which would spin a wet fulminate composition into the recess around the interior circumference of the head where it was allowed to dry. A perforated wad was then inserted to hold the priming in place within the rim. When the pistol's hammer nose struck the edge of the cartridge rim, the fulminate was crushed between the two interior surfaces and exploded.

This .22 caliber Smith & Wesson No. 1 cartridge was a success. No sooner had it been marketed, however, than B. Tyler Henry, plant manager for New Haven Arms Company, began experiments with a larger rifle cartridge suitable for the Volcanic tubular magazine weapons system which the company had acquired from the defunct Volcanic Arms Company in March 1857. The Volcanic .31 and .41 caliber primed bullet contained only enough fulminate and black powder to drive that projectile a short distance. However, Henry must have envisioned early on that the tubular magazine system was a practical device to feed a rimfire cartridge to the chamber, in fact tailor made for such application. He settled on .44 caliber, evidently the largest diameter he considered practical for an under-barrel tubular magazine. He patented improvements on the Volcanic in October 1860 including a double-strike firing pin and rimfire case extractor. Thus was born the Henry rifle and the .44 Henry rimfire cartridge.

The Henry sold well during the Civil War, although it never attained official acceptance by the military. With a 16-shot magazine, it could be fired 120 rounds in 340 seconds, including reloading time. The .44 rimfire cartridge it used was waterproof, breech sealing, and reasonably sure to fire. It lacked the weight of ball generally required by the military but the Spencer magazine rifle, the Ballard single shot, and many other systems were soon made to use the larger-caliber rimfire ammunition shown in the following pages. Ultimately, the .44 Henry was superceded by the Winchester Model 1866 after the war and it also saw military service around the world, particularly in Turkey. Percussion breech-loading rifles were modified to use rimfire ammunition and eventually muzzle-loading rifles were altered to breech-loading for rimfire ammunition.

Originally an American cartridge, the rimfire was adopted by Switzerland, Norway, Sweden, Denmark, and Austria for use in general issue infantry shoulder-fired arms.

References: "Winchester," Harold F. Williamson, 1952
"The American Cartridge,"Charles R. Suydam, 1960 (re-issued)
"Small Arms Ammunition at the International Exposition, Philadelphia, 1876, Col. Berkeley R. Lewis, 1972
"Civil War Guns," William B. Edwards, 1962
Correspondence and research records of Messrs. Victor v.B. Engel and George Kass

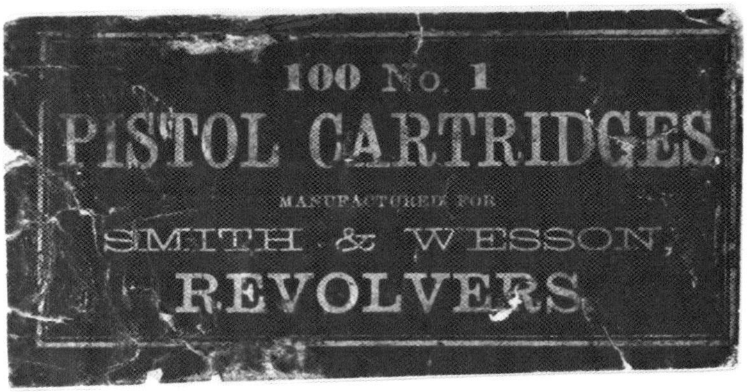
100 No. 1
PISTOL CARTRIDGES
MANUFACTURED FOR
SMITH & WESSON,
REVOLVERS

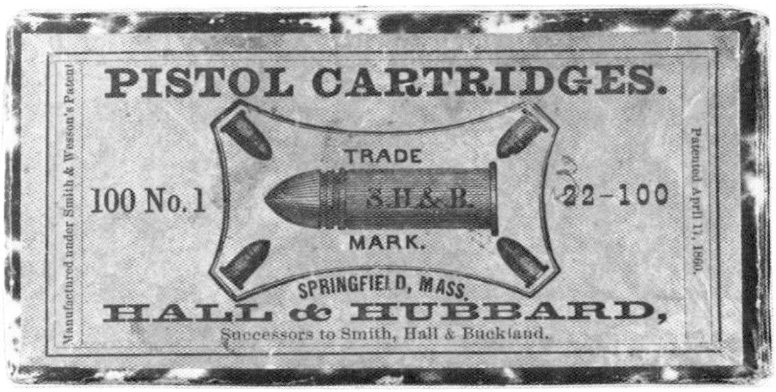
PISTOL CARTRIDGES.
TRADE
100 No. 1
S.H.& B.
22-100
MARK.
SPRINGFIELD, MASS.
HALL & HUBBARD,
Successors to Smith, Hall & Buckland.
Manufactured under Smith & Wesson's Patent
Patented April 17, 1860.

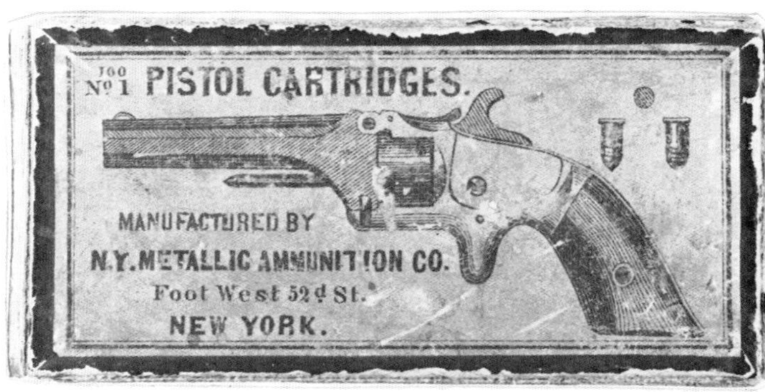
100 No 1 PISTOL CARTRIDGES.
MANUFACTURED BY
N.Y. METALLIC AMMUNITION CO.
Foot West 52d St.
NEW YORK.

RFR 1 – .44in (11.18mm) U.S. Henry, 1861, First Issue by New Haven Arms Co.

Description

Copper .820in case, .443in roundnosed lead bullet, no headstamp.

RFR 2 – .44in (11.18mm) Henry Flat, 1862

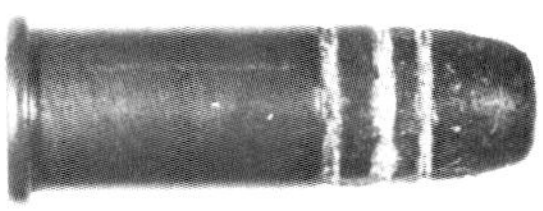

Copper .820in case, 443in flatnosed bullet, h.s. is a raised "H" in a circular depression. Found with one or two grooves.

RFR 3 – .44in (11.18mm) Henry, long case type

(a) Tinned copper .895in case, .445in flat-nosed bullet, no h.s.

(b) Copper .898in case, .438in flat-nosed bullet, no. h.s.

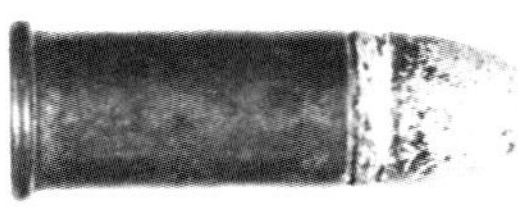

(c) Copper .898in case, mouth crimped over a cardboard, wad, no h.s. Blank for above.

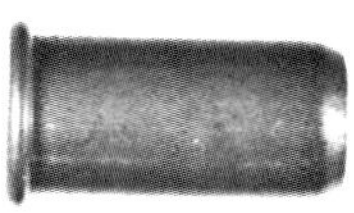

(d) Copper .900in case, .439in two-grooved blunt nosed bullet, h.s. is a raised "K" in depressed circle. Made by Keller & Co., Austria, for Turkish Government and used in the Model 1866 Winchester rifles purchased by that country.

Comments

The .44 Henry was made in large quantities by all of the principle ammunition makers of the United States and most of those in Europe. Consequently, a great variety exists in headstamps, case dimensions, and bullet styles. Most of the variations in the cartridge after 1865 must be considered commercial because the Henry and Model 1866 Winchester rifles became great favorites on the civilian market after the Civil War. Colt and Smith & Wesson revolvers were also chambered for it.

HENRY REPEATING RIFLE.

50 CARTRIDGES
FOR
HENRY'S REPEATING RIFLE
MANUFACTURED BY
ETHAN ALLEN & CO.,
WORCESTER, MASS.
No. 44—100.

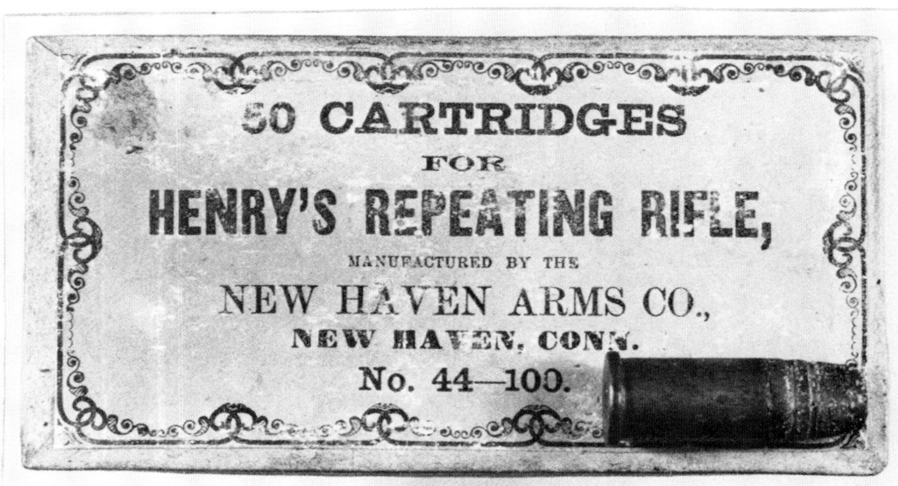
50 CARTRIDGES
FOR
HENRY'S REPEATING RIFLE,
MANUFACTURED BY THE
NEW HAVEN ARMS CO.,
NEW HAVEN, CONN.
No. 44—100.

B. Henry Tyler was granted the initial patent for his .44in rimfire magazine rifle on October 16, 1860. A direct descendent of the Hunt and Jennings rifles (see PBL 40), the tubular magazine was charged from the front by drawing up the coiled magazine spring into the muzzle secton of the tube beneath the barrel. The muzzle section was then swung around to the side, exposing the mouth of the magazine for insertion of 16 cartridges. The muzzle section was then turned back on the axis of the bore to close the magazine and the spring then compressed the column of cartridges to the rear. Bringing the lever down and forward withdrew the bolt which cocked the hammer. The spring threw a cartridge onto the carrier block and returning the operating lever to the closed position raised the carrier block and cartridge in line with the chamber. The cartridge was forced by the closing motion of the sliding breech bolt into the chamber and the rifle was ready to fire.

New Haven Arms Company under general management of Oliver F. Winchester sold 1731 Henry rifles to the Union and the remainder of some 10,000 made were bought by private individuals or units. It quickly became famous for its rapidity of fire, but the .44in, 210gr bullet lacked the striking force deemed necessary by military men of the day, so it did not achieve general acceptance. However, its successor, the Winchester Model 1866 using the same rimfire cartridge was sold to Mexico, Peru, Chile, France, and an important series of orders totalling over 50,000 Model 1866 muskets and carbines went to Turkey. The Model 1866 commercial models entrenched the Winchester lever action rifle as a favorite hunting arm among Americans and many models followed. The Model 94 is still made.

RFR 4 – .46in (11.68mm) Remington
Carbine, 1864

Description

Copper 1.11in case, .453in two-grooved pointed bullet, no h.s.

RFR 5 – .44in (11.18mm) Ballard
Carbine, 1861

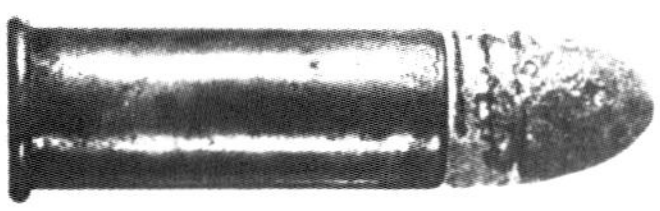

Copper 1.14in case, .443in two-grooved pointed bullet, no h.s.

Comments

The .46 caliber Remington carbine cartridge seems to have been relatively standardized as to case length and bullet style as made during the Civil War era. However, this does not seem to be the case with the .44 caliber Ballard. The specimen shown, although not identified with a particular maker, is believed to be the typical configuration of the early rounds used in the first Ballards as well as the Wesson and other carbines such as the Howard which appeared at that time but saw little or no military use. A typical commercial Ballard cartridge is included in the few such speciments illustrated at the end of this chapter on the rimfires. Others with shorter case and varying bullet style exist, several of which are shown by Colonel B. R. Lewis in "Small Arms and Ammunition in the United States Service."

Joseph Rider's November 15, 1864 patent on Leonard's Geiger's original idea for the "rolling block" action formed the basis for the "split breech" carbines made by Remington for the Union at Middletown, Conn. The specimen shown above is the .56-50 Spencer, one of 14,999 that saw service late in the war. A smaller-frame model chambered the .46 rimfire cartridge shown opposite. U.S. Ordnance bought 5,000 of these.

Charles H. Ballard patented the action on the carbine shown immediately below on November 5, 1861. Hammer and sear mechanisms were incorporated inside the dropping breechblock worked by the underlever trigger guard. Some 1,000 chambering the .44in rimfire cartridge shown opposite were bought by the U.S. Government in 1864. The specimen shown chambers the No. 56 Ballard (.56-56 Spencer), others were made for the longer .54in Ballard (which see). Ball & Williams of Worcester, Mass. manufactured these arms.

The second carbine below patented in October 1859 and November 1862 by Franklin Wesson (brother of Daniel B. Wesson of Smith & Wesson) was operated by placing the hammer at half cock and pulling the front trigger. This unlatched the barrel and allowed it to tip down, exposing the chamber for loading the .44in rimfire cartridge. The Union purchased only 151 during the war, but others were bought by state militias and individual units. B. Kittredge & Co., Cincinnati, Ohio made the Wesson. (Milwaukee Public Musuem photos)

RFR 6 – .36in (9.14mm) Unknown Carbine, 1860.

Description

Copper .943in case necked and crimped into a four-groove, .372in lead bullet. No h.s.

RFR 7 – .56-56in (13.21mm) Spencer Rifle and Carbine, ca. 1860

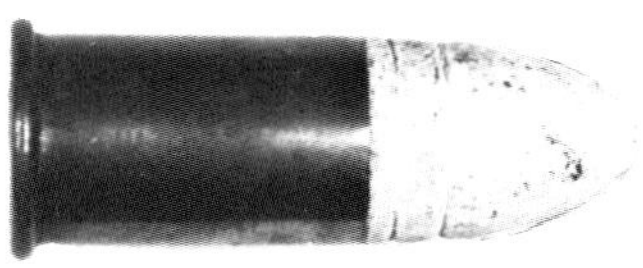

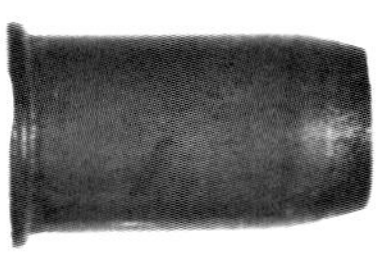

(a) Copper .89in case, .542in pointed lead bullet with two grease grooves. No. h.s. The original Navy and Infantry cartridge made by Crittenden & Tibbals, South Coventry, Conn.

(b) Copper .89in case, .547in wood bullet. Dummy for above. No h.s.

(c) Copper .097in case, crimped over a varnish-covered black powder charge. Blank for above. No h.s.

(d and e) Cutaways of RFR 7 with 450 and 440 gr (29.16 and 28.51 gm) bullets. No h.s. Displayed by U.S. Ordnance at the 1876 Philadelphia Exposition.

Comments

Christopher Spencer's first small-frame carbines made under his 1860 patent were chambered for the .44 Henry rimfire and an unknown .36 rimfire. The specimen at RFR 6 is shown because it exists, is similar in bullet style and general appearance to a later .56-46 Spencer commercial cartridge, and so may be one of the early Spencer experimentals. U.S. military authorities would not approve of .36 or .44 caliber so the light model was dropped and a heavier arm built to chamber the .56-56. This remained the standard cartridge for the Spencer rifle and carbine throughout the Civil War. Spencer cartridge designation reflects the diameter of the cartridge head forward of the rim and at the case mouth. The rifle and carbine were actually .52 caliber (13.21mm).

RFR 8 – .56-50 (12.70mm) Spencer Carbine, 1864

Description

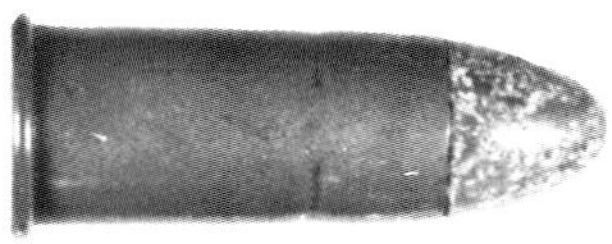

(a) Copper 1.17in case crimped over the lubricating grooves on a .501in lead bullet and showing four crimping stabs in side of case. Headstamped "50 CAL" indicating manufacture at Springfield or Frankford Arsenals.

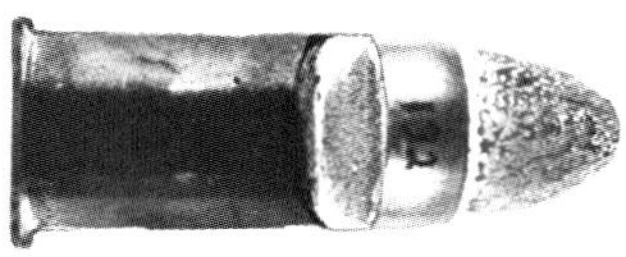

(b) Cutaway specimen of above. No h.s. Displayed by U.S. Ordnance at the 1876 Philadelphia Exposition.

RFR 9 – .56-52 (12.70mm) Spencer, ca. 1866

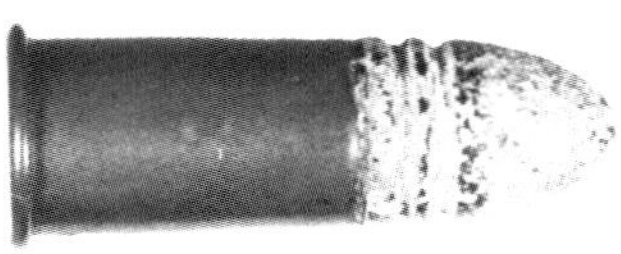

(a) Copper .95in case slightly necked to hold a .534in two-groove bullet. Case headstamped "CDL" by C. D. Leet, Springfield, Mass.

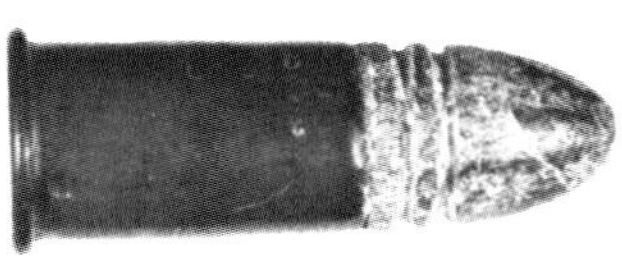

(b) Copper .96in case with tapered crimp over a two-groove bullet. No headstamp.

Comments

The .56-50 was the result of efforts begun in 1863 at Springfield Arsenal to design a new cartridge for this and other rimfire carbines in use by Union forces. By the fall 1864, the cartridge had been perfected and was ordered adopted. However, no Spencer carbines chambered for this cartridge were finished in time for use during the Civil War. Ultimately, other rimfire shoulder arms chambered for similar cartridges were altered or manufactured new to accept the newly-standard military round, the .56-50.

The .56-52 has often been labeled the second in the Spencer series. However, Colonel Lewis found no evidence that it was used during the Civil War. Edwards states that it was a Spencer creation after the .56-50 was developed. In any case, the .56-52 chambers and fires in a .56-50 carbine with .50 caliber barrel. The essential difference between the two cartridges is the case length. The specimens are shown because they were usable in a military shoulder arm, though probably saw most use after the Civil War as a hunting cartridge.

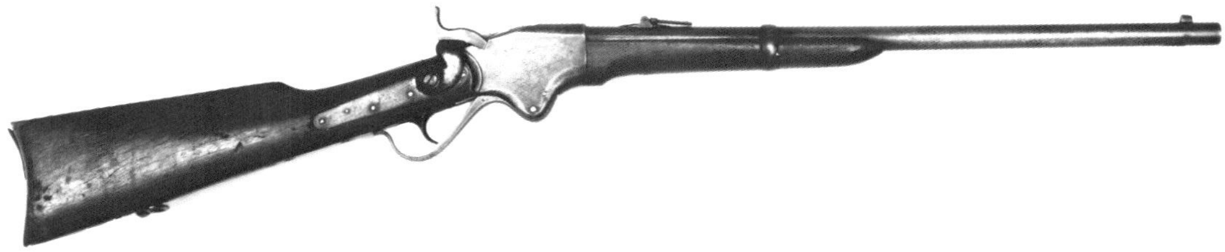

Christopher Spencer's carbine patented March 6, 1860 featured a magazine within the butt stock which was charged through a trap in the butt plate. Loading one cartridge at a time was slow until the appearance of the Blakeslee patent cartridge box which held ten tin tubes containing seven cartridges each, the magazine capacity. This device permitted loading all seven at once. Pulling down and forward on the underlever trigger guard rotated the breechblock down and back, ejecting the spent cartridge case, and feeding a fresh cartridge onto the block. Returning the lever to its closed position chambered the cartridge and locked the breech. The Stabler cut-off mechanism developed for later models limited the down throw of the operating lever so that the magazine capacity could be maintained while the gun was loaded with single rounds by hand. The carbine was cocked by hand for each shot. The Spencer Rifle Company of Boston, Mass. manufactured this arm, though Burnside Rifle Co., Providence, R.I. made some of the .56-50's. One of the most important arms used by the Union forces, a total of 94,196 were made for wartime use.

The carbine illustrated above, patented on December 22, 1863 by William C. Palmer is shown because it is one of those made to chamber the .56-50 Spencer cartridge. It also had the distinction of being the first bolt action shoulder arm using a metallic cartridge which was purchased by the U.S. Government. E. G. Lamson & Co., Windsor, Vt., the maker, manufactured at least 1,001 bearing the date 1865 and sold them to the Union. However, they never saw service. Lifting up on the handle situated on the left rear side of the bolt unlocked the action and allowed the bolt to be drawn to the rear, exposing the chamber for loading. The conventional hammer struck an exposed part of the cartridge rim to fire. An extractor hook was located in the top of the bolt face. (Milwaukee Public Museum photos)

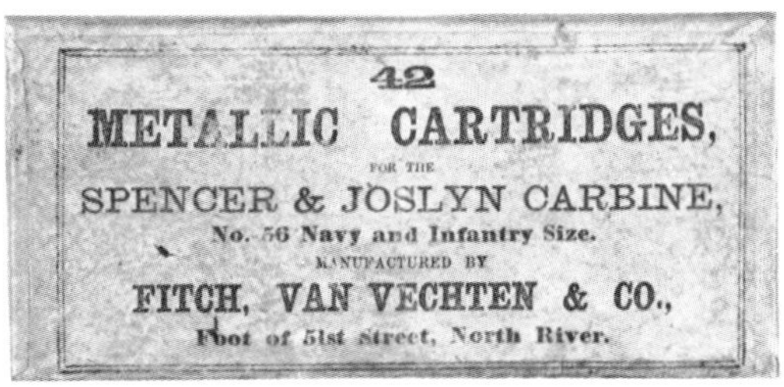

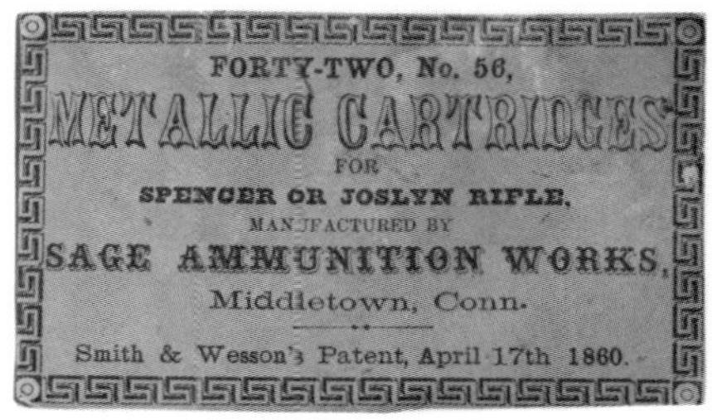

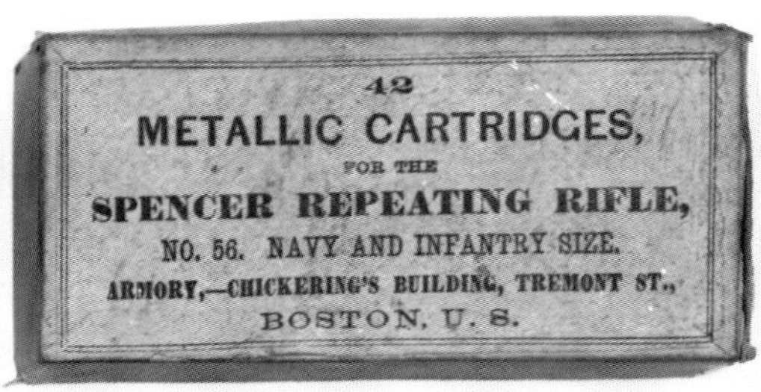

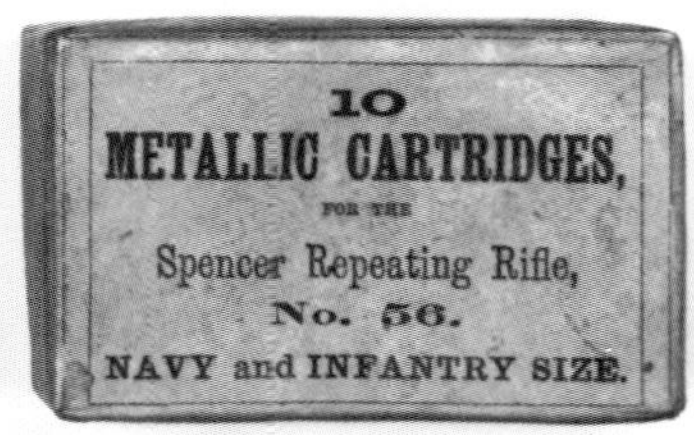

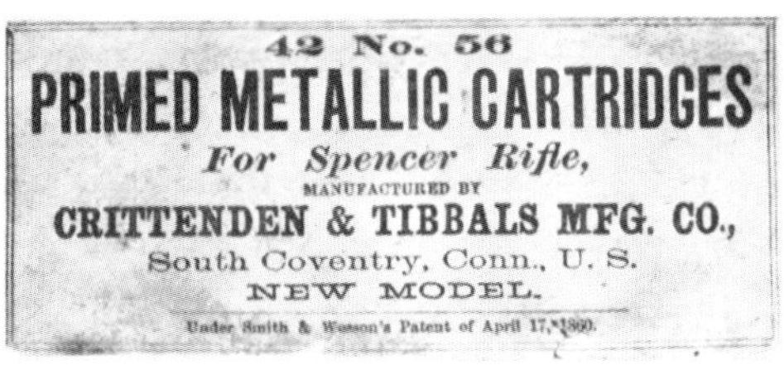

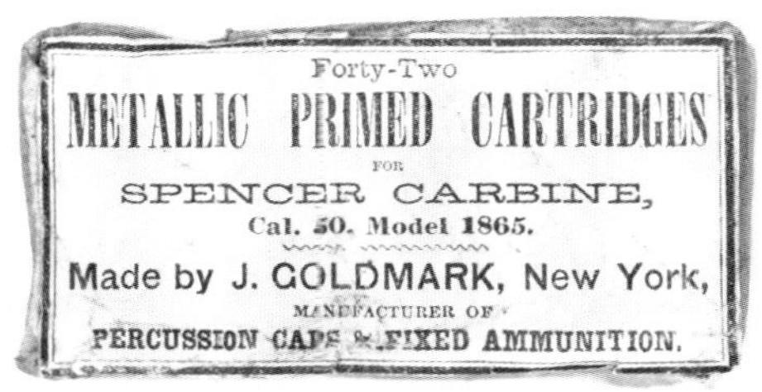

See appendix section for more Spencer and other rimfire rifle box labels.

RFR 10 – .54in (13.72mm) U.S. Ballard,
ca. 1861-1865

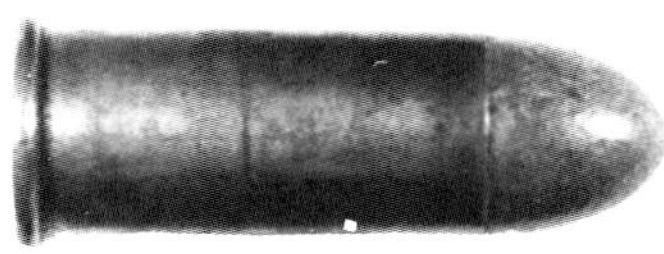

Description

(a) Copper solid drawn 1.22in case, .540in lead bullet, no grooves. No h.s. This is the Ballard Old Model rifle load.

(b) Copper 1.22in case, slight indentation of the head indicating the Smith & Wesson patent wad is in the interior of the base. No h.s. but probably made by Crittenden & Tibbals. This is the shorter-bullet Ballard carbine load.

Comments

The Ballard rifle and carbine cartridges in .54 caliber are recognized by their considerably longer case than the Spencer and the distinctive rounded-point lead bullet with no lubricating grooves. In addition, the linen patch covered by the Elliott patent used on the Ballard bullet is often visible at the case mouth. A case variation not shown here is the Smith & Wesson patent two-piece with the separate head crimped over the slightly flanged base of the case. This two-piece fold is evident from a series of small creases or folds all the way around the rim.

It should be noted that the very early Ballard carbines were chambered for a cartridge identical in dimensions to the .56-56 Spencer, and would fire that cartridge. Some sources refer to a No. .56 Ballard cartridge. The second page of rimfire box labels in the appendix section shows a C. D. Leet box label reading–"25 No. 56 METALLIC CARTRIDGES For Ballard's Carbines." A specimen from that box is identical in dimensions to the Navy and Infantry size .56-56 Spencer round. These cartridges were either ordered with such a label or C. D. Leet was canny enough to label them that way for those that owned or were issued such carbines. Other box labels shown in the appendix identify the cartridge for use in Spencer's rifle and Joslyn's carbine, and one identifies it as for the Joslyn carbine. They were all the same cartridge. The .54 Ballard and the .56-56 Spencer (No. 56 Ballard or Joslyn) were loaded with .54 caliber bullets, but the long Ballard cartridge will not chamber in a carbine made to use the shorter-chamber Spencer-type cartridge.

U.S. Ordnance Department purchased a small quantity of the early Ballard .54 caliber rifle for tests. The principle order for them numbering some 15,000 went to the Kentucky State Militia. The rifle illustrated first above is .54 caliber. The second, a different model, is .46 rimfire. Both were made with the extractor arm working in a slot in the bottom of the forearm which was operated by grasping the knob visible on the second rifle and pulling it to the rear. This device evidently has been broken off or lost from the first specimen shown.

The C. H. Ballard patent of November 5, 1861 illustrates an extractor working directly from the operating lever. However, this apparently did not work well and only 100 or so were made with that mechanism. (Milwaukee Public Museum photos)

RFR 11 – .52in (13.21mm) U.S. Sharps & Hankins Rifle and Carbine, ca. 1861-1867

Description

(a) Copper 1.18in case, .547in round-nosed bullet with one band, no h.s. This is the cartridge for the rifle.

(b) Copper 1.15in case, .554in flat-nosed carbine bullet, one groove, no h.s.

(c) Copper 1.02in case, .536in flat-nosed carbine bullet, two grooves showing, no h.s. A variant load, probably earlier than the specimen shown at (b), but same overall length.

Comments

The Sharps & Hankins rifle cartridge is also found with a single groove on the round-nosed bullet. The carbine cartridge is also distinguishable from the rifle round by a slight crimping for .10in of the case mouth. Presumably these cartridges were made by Sharps & Hankins in Philadelphia. Specimens also exist made by Ethan Allen and Crittenden & Tibbals.

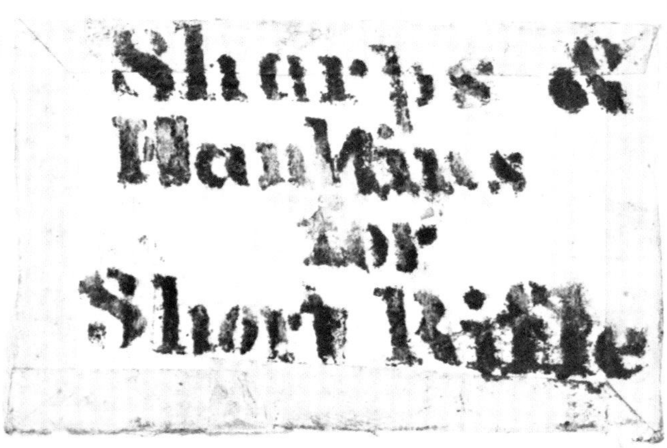

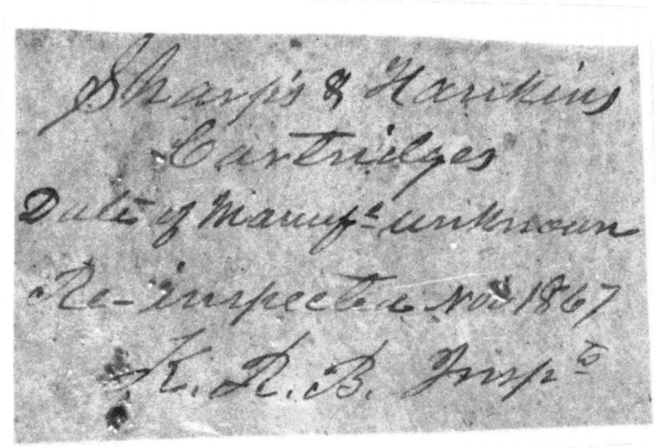

Christian Sharps suspended his relationship with the Sharps Rifle Company in the early 1850's and moved to Philadelphia, Pa. where he resumed experiments with other firearms actions. He patented the action for the above rifle on July 9, 1861 and subsequently formed a partnership with William C. Hankins to manufacture it. This model, designed for rimfire cartridges, differed radically from his 1848 dropping block system.

When the underlever trigger guard was pushed down and forward, an operating rod pinned to it and to the rear bottom of the barrel slid the barrel assembly forward in its track in the extended part of the frame which served as a forearm. A cartridge could then be inserted in the chamber and drawing the lever back brought the barrel face against the standing breech. A carbine model made for the Navy was first issued with tinned barrel, later with a leather cover around the barrel.

Sharps & Hankins manufactured ammunition for their own rifles and at least the .56-56 Spencer rimfire, also adapted to the Joslyn carbine (See appendix for box label. Milwaukee Public Museum photos).

RFR 12 – .54in 13.72in) U.S. Joslyn
Rifle and Carbine, ca. 1861

Description

Copper .89in case, 542in two-grooved lead bullet, no h.s. Made by C. D. Leet, Springfield, Mass. under the Smith & Wesson patent.

RFR 13 – .58in (14.73mm) U.S. Joslyn
Carbine, ca. 1862(?)

Copper .750in case, .598in bullet with no grooves showing, no h.s.

Comments

Existing box labels show that C. D. Leet made the No. 56 Joslyn cartridge (referred to as the .54 caliber above because that was the caliber of the arm and the diameter of the bullet) and identified it as specifically for that arm. However, the cartridge is essentially the same as that made for the .56-56 Spencer, a fact confirmed by the Sharps & Hankins box label. The .58in Joslyn cartridge, and carbine, are much less common than the .54in. Why the company made such a short-range round is unclear, nor is there positive evidence that it was used during the Civil War or later for military purposes.

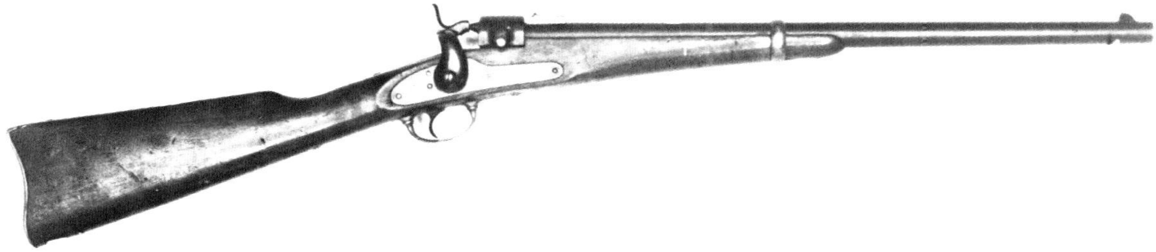

Benjamin F. Joslyn's first firearm patent of August 28, 1855 covered a breech opening system with the operating lever lying along the top of the tang which tipped up to lift the breechblock and expose the chamber. His later patent of October 8, 1861 described the system shown above. The breechblock was hinged on the left side of the frame. Lifting up on the lever on the right side swung the block up and counter-clockwise to expose the chamber. Opening the action cammed an ejector back to draw out the spent cartridge case. The block was made with a ring which fit into a groove turned on the rear portion of the barrel to lock the action when closed. Over 11,000 of these carbines manufactured by Joslyn Firearms Co., Stonington, Conn. were delivered to U.S. Ordnance department by February 25, 1865. (Milwaukee Public Museum photo)

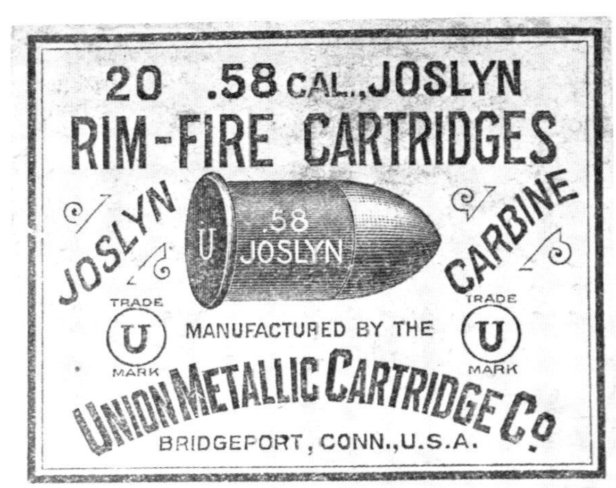

Description

RFR 14 — .50in (12.70mm) U.S. Ball Carbine, 1864

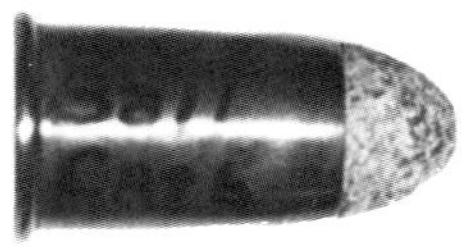

Copper .859in case, .456 bullet with case crimped well over the widest diameter of the bullet. No h.s.

RFR 15 — .50in (12.70mm) U.S. Warner Carbine, ca. 1864

Copper .850in case, .515in two-grooved pointed bullet, no h.s. This cartridge is smaller in overall dimensions than the military Spencer cartridges.

Comments

The Ball and Warner cartridges are examples of proprietary loads which the inventors of these arms developed. The Ball was seriously under-powered for practical military use and the Warner was so similar to the .56-50 Spencer tha t both were soon dropped and the arms made to accept the U.S.-developed 1864 Spencer cartridge.

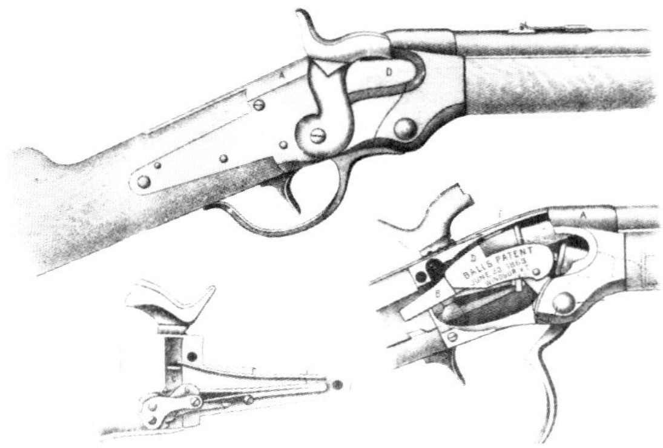

The Ball & Lamson carbine patented August 16 and December 6, 1864 resembled the Spencer in operation but the magazine tube was situated within the forearm. Nine of the original short cartridges were held under spring pressure and fed one at a time onto the block when the operating lever was drawn down and forward. Returning the lever to its closed position moved a cartridge into the chamber ahead of the block. The hammer nose struck an exposed part of the cartridge rim to fire. The Ball & Lamson could also be worked as a single loader. U.S. Ordnance records reflect that 1,002 of these carbines were purchased during the Civil War but it is doubtful they saw service. E.G. Lamson & Co., Windsor, Vermont, manufactured these carbines.

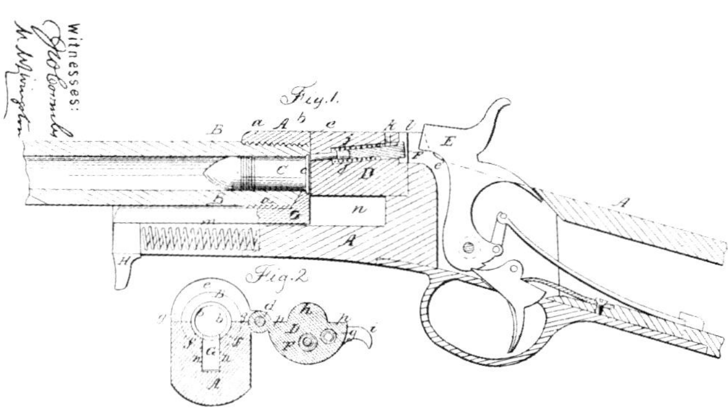

The action shown above patented by J. Warner on February 23, 1864 was used in the brass-framed carbine manufactured by Greene Rifle Works, Worcester, Mass. The breechblock was hinged to the right side of the frame. To operate, the hammer was brought to half cock and the iron catch to the left of the hammer depressed. The block could then be swung open clockwise. Extraction of the empty case was accomplished by a lever under the frame which was operated by hand. However, the gun had to be turned over to let the case drop out. Despite its late appearance, 4,001 of these Warners were bought by the Union, and most were later converted to .56-50 Spencer. (Illustrations from Schmidt's "Les Armes A Feu Portatives," 1877, and the original patent)

RFR 16 – .54in (13.72mm) U.S. Allen & Brand, ca. 1864

Description

Copper .89in case, .556in bullet, no h.s.

RFR 17 – .50in (12.7mm) U.S. Jenks Carbine, ca. 1864

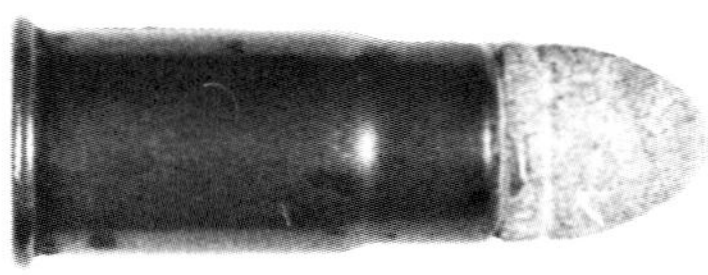

Copper 1.26in case, .539in swelled lead bullet with one shallow groove. Case is necked .30in below the mouth. No h.s.

Comments

The Allen & Brand cartridge is the same as the .56-56 Spencer, but the bullet is 473gr (30.65gm), or 23gr larger than the heaviest Spencer bullet. The difference can be detected in the unusual length of the Allen & Brand bullet. Col. Lewis lists an Allen & Brand .58in cartridge loaded with three buck, wad, and ball. This cartridge is represented in at least two collections in the U.S.

The Jenks cartridge must have been one of the last proprietary loads specifically designed for a rimfire carbine. The patent is dated in 1864 but it did not appear for trials until early 1865. The Jenks was among many that were not adopted during the Springfield Arsenal examinations of various weapons at that time and the cartridge undoubtedly never survived this experimental stage. The first positive identification of this previously unknown cartridge was made in 1979 as a result of discovery and publication of a March 16, 1868 paper prepared by Capt. J. B. O'Hea, British Army, entitled "Cartridges for Breech-loading Small-Arms, and The Best Form of Projectile." This paper in its entirety appears in the March-April 1979 issue of The International Cartridge Collector, courtesy of the editor, John A. Belton. The Jenks cartridge is illustrated therein.

35,989 C. C. BRAND. July 29 1862

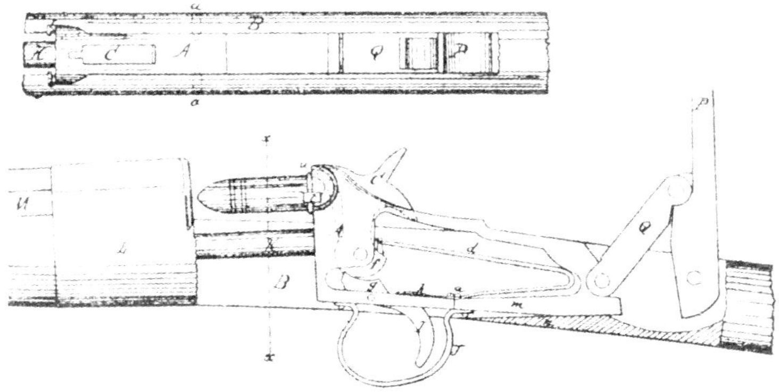

The 1862-patented Brand action provided for a breech pin sliding in a recess in the stock, so that the frame and lock could be pulled away from the barrel assembly by lifting up on the operating lever lying on the tang. The standing breech was recessed to accept the head of a cartridge which was inserted and held there by a hook until action and barrel assembly were brought together to chamber the cartridge ready for firing. Brand patented a slightly different arrangement, though operating on the same system, on April 28, 1863. In "The Breech-loader in The Service," 1933, Claud E. Fuller illustrates the "#1 Allen" made with this action as one of the arms entered in the early 1865 trials held at Springfield Arsenal.

74,760. B. H. JENKS. Feb. 25, 1868.

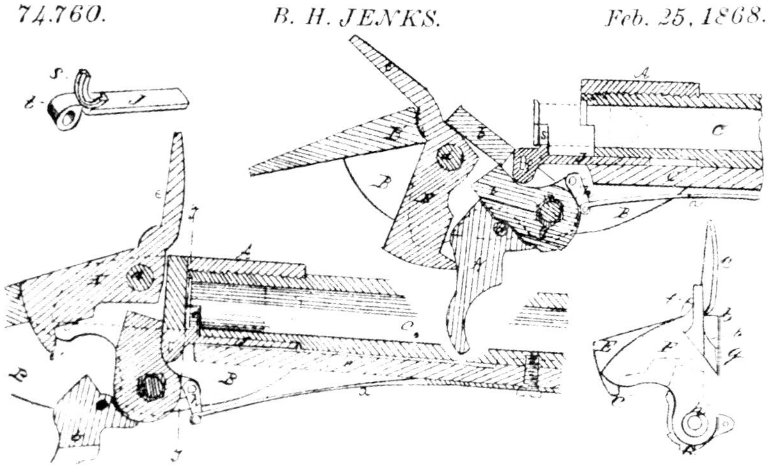

The Jenks carbine was also entered in these trials. The patent drawing reflects a rotating breechblock operated by an upright lever arrangement. The descripton reads, in part—". . . Combining a vibrating slotted breech-piece D, a vibrating locking-latch E, and a hammer F, in such manner that these parts are confined between the cheeks B, of the frame, and the hammer is on the outside of the breech-piece . . . ." Schmidt's illustration in "Handfeuerwaffen," 1875, shows the upright hammer and an operating handle to the right of it which opened and closed the action. (Illustrations from the original patent drawings)

RFR 18 – .45-45in (11.43mm) U.S. Peabody Rifle, ca. 1862-1872

Description

Copper 1.35in case, .470in lead bullet, no h.s.

RFR 19 – .45-60in (11.43mm) U.S. Peabody Rifle, ca. 1862-1872

Copper 1.485in case, .470in bullet, no h.s.

RFR 20 – .45-70in (11.43mm) U.S. Peabody Rifle, ca. 1862-1872

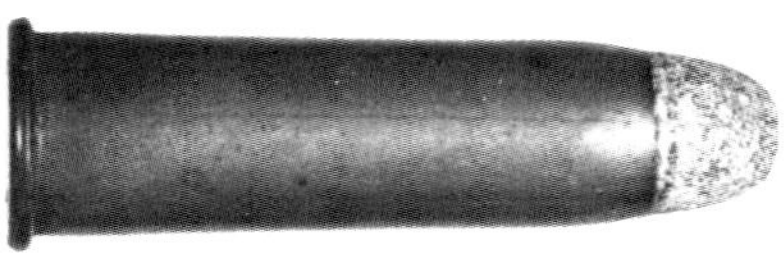

Copper 1.702in case, .470in bullet, no h.s.

RFR 21 – .45-? (11.43mm) U.S. Peabody Rifle, ca. 1862-1872

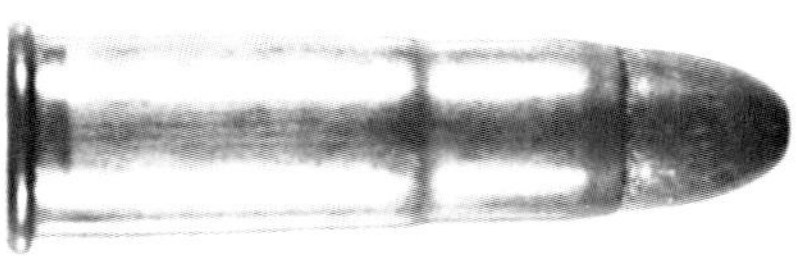

Copper 1.613in case, .470in bullet, no h.s. Slight bottle necked case.

Comments

This is believed to be the complete series of experimental .45in cartridges for the early Peabody single shot rimfire rifles and carbines. Switzerland tried out the Peabody system extensively in the latter part of the 1860's before finally settling on the 10.4mm rimfire. At least RFR 21 has been determined to be one of the Swiss-originated case types but probably manufactured in the U.S. for them. A distinctive identifying feature of these Peabody cartridges is the heavy crimp of the case mouth over the largest diameter of the bullet. The first number in the designation of these cartridges identifies the nominal bore diameter of the rifle, the second reflects the weight of the powder charge.

RFR 22 — .50-45in (12.7mm) U.S. Peabody Rifle, ca. 1862-1872

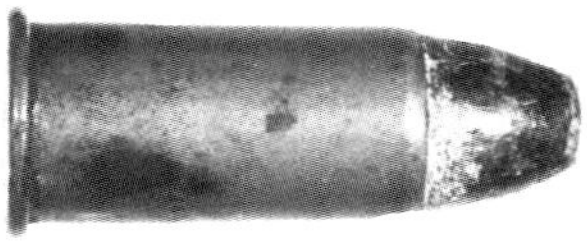

Description

Copper 1.162in case, .520in semi-pointed lead bullet, no h.s.

RFR 23 — .50-50in (12.7mm) U.S. Peabody Rifle, ca. 1862-1872

Copper 1.364in case, .520in semi-pointed wood bullet. Headstamp is an impressed "BLANK."

RFR 24 — .50-50in (12.7mm) U.S. Peabody Rifle, ca. 1865-1877

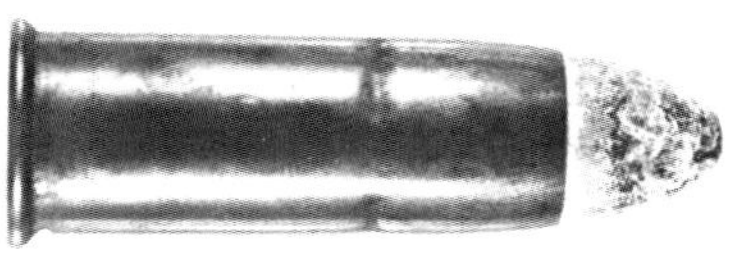

Copper 1.477in case, .520in semi-pointed lead bullet. Headstamp is impressed "50 CAL."

RFR 25 — .50-70in (12.70mm) U.S. Peabody Rifle, ca.1862-1872

Copper 1.665in case, .520in semi-pointed lead bullet, no. h.s.

Comments

Among the four known .50 caliber Peabody rimfire cartridges, only the .50-60 was adopted as a military load. Canada ordered 5,000 rifles in this caliber in 1866, but only 3,000 were delivered. U.S. arsenals also experimented with the load as evidenced by the head stamp on the specimen shown. The .50 caliber series cartridges exhibit the same rounded case mouth crimp as seen in the .45 caliber series, though not quite as pronounced because of the larger size bullet. Bullets in the first three are noticeably more pointed than the .45 caliber series. As in the previous group, cartridge designation stems from the nominal bore diameter of the rifle and the weight of the powder charge.

RFR 26 – .45-70in (12.7mm) U.S. Peabody Rifle, ca. 1862-1872

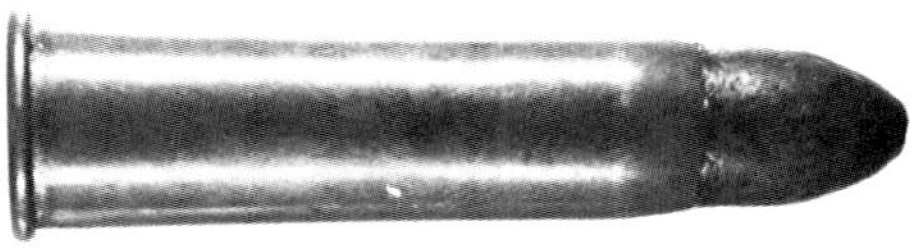

Description

Copper 1.801in case, .470in lead bullet, no h.s.

RFR 27 – .50-60in (12.7mm) U.S, Peabody Rifle, ca. 1865-1877

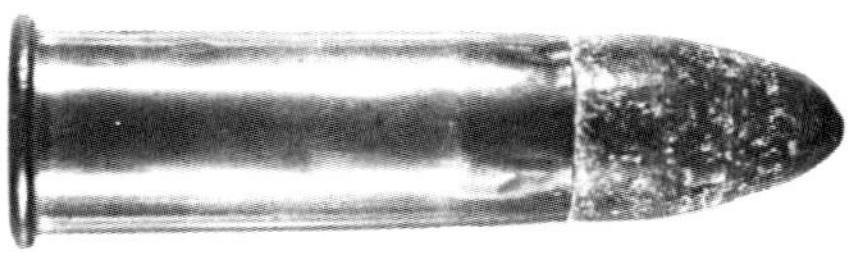

Copper 1.463in case, .520in lead bullet, no h.s.

RFR 28 – .50-60in (12.7mm) U.S. Peabody Rifle, ca. 1865-1877

Copper 1.471in case, .520in lead bullet. Headstamp is an impressed "50 CAL."

Comments

The three specimens shown here are proof loads with heavier charge and/or heavier bullet than normal. The last .50-60 shown was U.S. arsenal made.

The loads were all relatively mild compared to what the Peabody rifle could actually withstand. During the early 1865 trials of various breech-loading systems by the United States Army Board convened at Springfield Arsenal, the Peabody came out among the best of four during preliminary tests. These four were ultimately fired with 60gr (3.89gm) of powder behind three bullets each weighing 450gr (29.2gm). Finally, the charges were increased to 80gr (5.18gm) powder and five bullets of the above weight. Only the Peabody rifle withstood this test. The final test of the Peabody was 120gr (7.78gm) powder with the same five projectiles and the action incurred no damage.

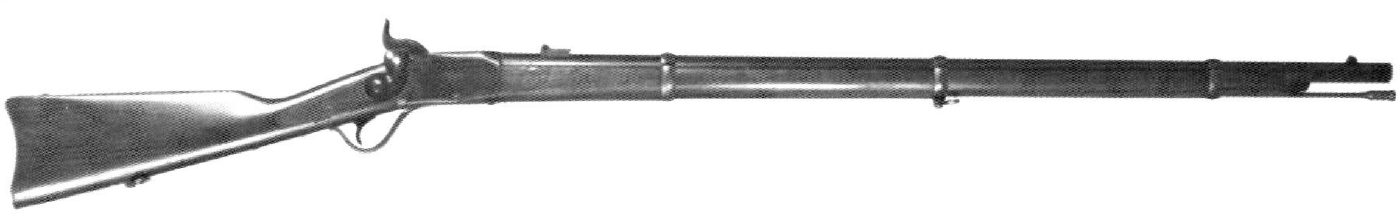

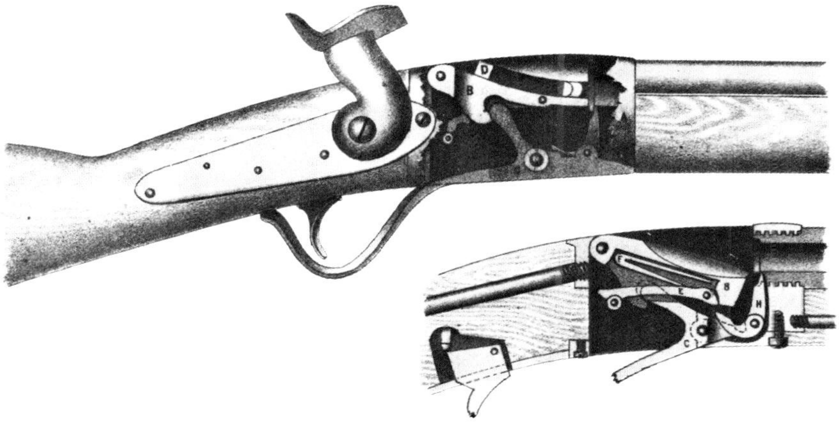

Henry O. Peabody's breech-loading rimfire action patented July 22, 1862 described a breechblock rotating at its rear end on a hinge pin which allowed the block to tilt downward exposing the chamber when the trigger guard underlever was thrown down and forward. The block in turn depressed a two-piece lever arrangement which extracted the fired case from the chamber. The fresh cartridge was inserted along the race in the top of the block into the chamber. When the operating lever was returned to its closed position, the breechblock was braced between the face of the chamber and the rear of the frame.

Peabody asked the U.S. Ordnance department to test his rifle in 1862 and trials conducted in 1863 were successful. However, Providence Tool Co., Providence, R.I., the makers of the Peabody, received no order for arms. Delays occurred while the Ordnance department deliberated over a common caliber for all small arms and then decided to seek a standard breech-loader. Finally, the Peabody showed its superiority in strength over all others during the early 1865 trials. Six of the rifles were made at Springfield Armory, but the war was over by then and this brought to a halt further arms acquisitions by the Union.

The Peabody rimfire fared better with other governments. Canada purchased 3,000 of the 50-60 model in 1866 and Switzerland ordered 15,000 chambered for its 10.4mm (.41in) rimfire cartridge in 1867. Other nations tested and approved of the system but did not purchase the Peabody in rimfire, presumably because superior centerfire cartridges were becoming available. The last military order for the .50-60 rimfire went to South Carolina State Militia in 1877 – a total of 350 from the factory's leftover stock.

Providence Tool Co. began making centerfire models of the rifle shortly after the Civil War and sold quantities to other state militias and foreign governments. The hammerless version of the rifle called the Martini-Henry which the British government adopted without credit to Peabody was also made by Providence Tool Co. and sold to Roumania and Turkey. The last was a huge order, but this story belongs in Volume II devoted to the world's centerfire military cartridges. (Illustration from Schmidt's "Les Armes A Feu Portatives," 1877)

RFR 29 – .50-60 (12.70mm) U.S. Joslyn,
1865 Conversion of Rifle-Musket

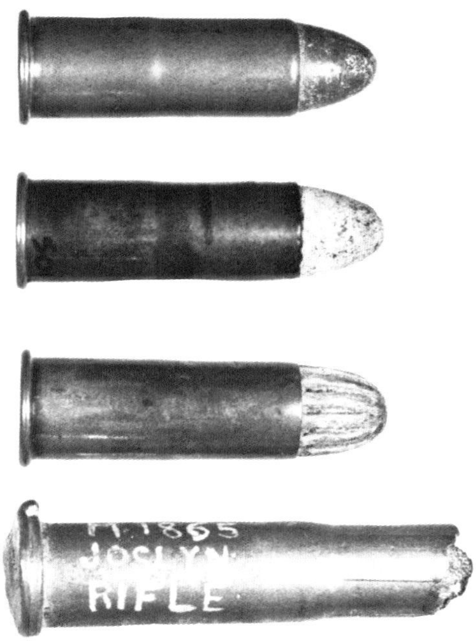

Description

(a) Copper 1.47 in case, .520in bullet, slight neck in case .70in from mouth, no h.s.

(b) Same as above, except for slightly longer bullet. No hs.

(c) Same as above but with odd fluted bullet, evidently an experimental idea. Headstamp is “CDL” (C. D. Leet).

Chamber cast of rifle shown opposite reflecting the slight necking also visible on the cartridges.

Comments

These cartridges are so close to being the 50-60 Peabody that they could be the same, except that the slight necking well down the case body is absent on the 50-60 Peabody. Comparison of the relative dimensions suggests that the straight-case Peabody round would chamber in this Joslyn conversion but these shown here might not chamber in the Peabody rifle.

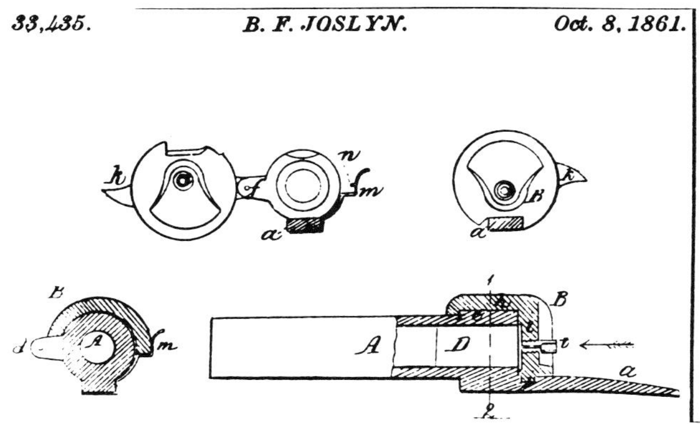

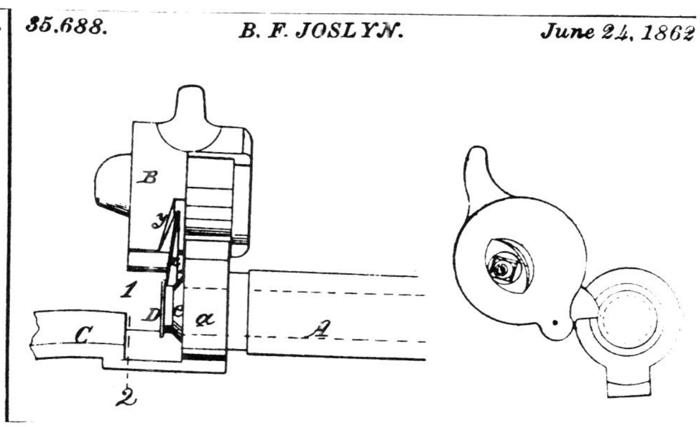

In 1865, Springfield Armory fabricated 3007 .50 caliber rifles using the Joslyn action patented October 8, 1861, the same one used on the Civil War carbine previously shown. These have been described as conversions of muzzle-loaders. However, since there was no .50 caliber muzzle-loading rifle to convert, these were furnished with new barrels and stocks, and only the lock mechanisms were obtained from obsolete muzzle-loading rifles.

During tests, a 45-gr (2.92gm) load was tried, which would suggest that a cartridge similar, if not identical, to the Peabody .50-45 was used. The charge was subsequently increased to 60 gr (3.89gm) as contained in the specimens illustrated opposite. The Springfield Joslyn rifle was just one of the several actions experimented with at that time which lost out to the Allin "trapdoor" mechanism. (Illustrations from the original patents)

RFR 30 – .52-70 (13.21mm) U.S. Sharps for Converted Percussion Rifle and Carbine, ca 1865

Description

(a) Copper 1.47in case, .529in two-groove lead bullet, no h.s.

RFR 30(a) – .52-(?)(13.21mm) Unknown, presumed U.S.-made

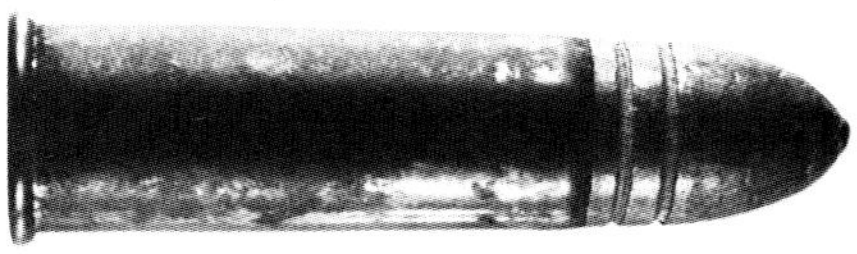

Copper 1.541in case, .528in lead bullet with two grooves. No h.s.

RFR 31 – .54in (13.72mm) U.S. Starr for converted Percussion Carbine, ca. 1867

Copper 1.03 in case, .540in two-groove pointed lead bullet, no h.s.

Comments

Frank Sellers reports in "Sharps Firearms" that the company experimented with converting their percussion breech-loader to rimfire in 1861. Navy trials conducted in July of that year used .56 caliber rimfire cartridges, 300 with copper cases and 200 with brass cases. These were probably the No. 56 Spencer cartridge (.56-56), since the Sharps was .52 caliber, the same as the Spencer.

The .52-70 was developed after the war for rifles that chambered the .52 caliber paper or linen ammunition. However, bore diameters were not consistent so the decision was made to line the chambers or re-line the barrel altogether to take the rimfire .50-70, later the centerfire .50-70 round. The .52-70, and a companion .52-70 centerfire were then dropped. The second cartridge shown above, may be related to those experiments though its rim, head, and case length are slightly larger than the .52-70.

The Starr carbine, with its tilting back percussion breech-loading action, was modified to replace the percussion nipple with a rim firing pin housed in the breech. The cartridge has the same rim and head diameters as the .56-56 Spencer and that cartridge was also used in the Starr rimfire. However, the case of the Starr cartridge shown above is considerably longer than that of the Spencer round.

Conversion of the Sharps straight breech rifle and carbine required a new breechblock with rim firing pin assembly and associated parts. The percussion primer feeding system was removed from the lock plate and the hammer nose was filled and reshaped to strike the firing pin rather than percussion nipple. Several other modifications were necessary but most of these were not visible. Therefore, the Model 1867 and 1868 conversions looked very much like their predecessor described in Chapter 7 that used paper or linen percussion ammunition.

The Starr carbine shown above took on a distinctly different appearance after conversion from percussion to rimfire. A new hammer mounted vertically on the lockplate replaced the long, curved type used on the percussion model.

Others of the Civil War era percussion carbines were converted to rimfire cartridge experimentally but the Sharps and Starr are the only two identified during this study for which distinctly new rimfire cartridges were developed—those two illustrated opposite.

According to Sellers, Sharps Rifle Company converted 32,184 rifles and carbines to rimfire for the U.S. Government, most of them carbines. He estimates that some 44,000 altogether were converted, including those altered by private individuals or firms. All were the straight breech type as the slant breech action was not adaptable to conversion. An illustration of a rimfire conversion breechblock is illustrated in his chapter on the conversion models. However, nowhere could be found in any reference a breakdown on the number of rimfire Sharps rifles and carbines converted from percussion. Probably there were a few of the Model 1867 since the centerfire .50-70 cartridge supplanted the rimfire. (Milwaukee Public Museum photo)

Reference: "Sharps Firearms," Frank Sellers, 1978.

RFR 32 – .58in (14.73mm) U.S. Mont-Storm, 1858 Transformed Rifle-Musket

Description

Copper 1.23in case, .602in grooveless, pointed bullet, no h.s. The largest of the U.S. so-called .58 caliber rimfires.

RFR 33 – .69in (17.53mm) U.S. Roberts, 1866 Transformed Rifle-Musket

Copper .748in case, .724in flat-nosed, shallow-grooved lead bullet. No h.s., but made by C.D. Leet.

RFR 34 – .69in (17.53mm) U.S. Crittenden & Tibbals, ca 1866, for Transformed Rifle-Musket

Copper 1.07in case, .713in grooveless pointed bullet, no h.s.

Comments

The rimfire cartridge had not long been perfected in the United States when military authorities began considering the large quantities of .58 and .69 caliber rifle-muskets on hand which could be converted to use such breech-loading ammunition. The first cartridge shown above was made to chamber in the Mont-Storm action originally patented in 1856, modified to use rimfire cartridges (see Percussion Breech-loader 14, Chapter 7). The huge .69 caliber Roberts and Crittenden & Tibbals made in experimental quantities only were obsolete before they were ever manufactured because technical advancements in firearms made during the Civil War did away with massed troops tactics, the only circumstances in which such short-range ammunition would be useful.

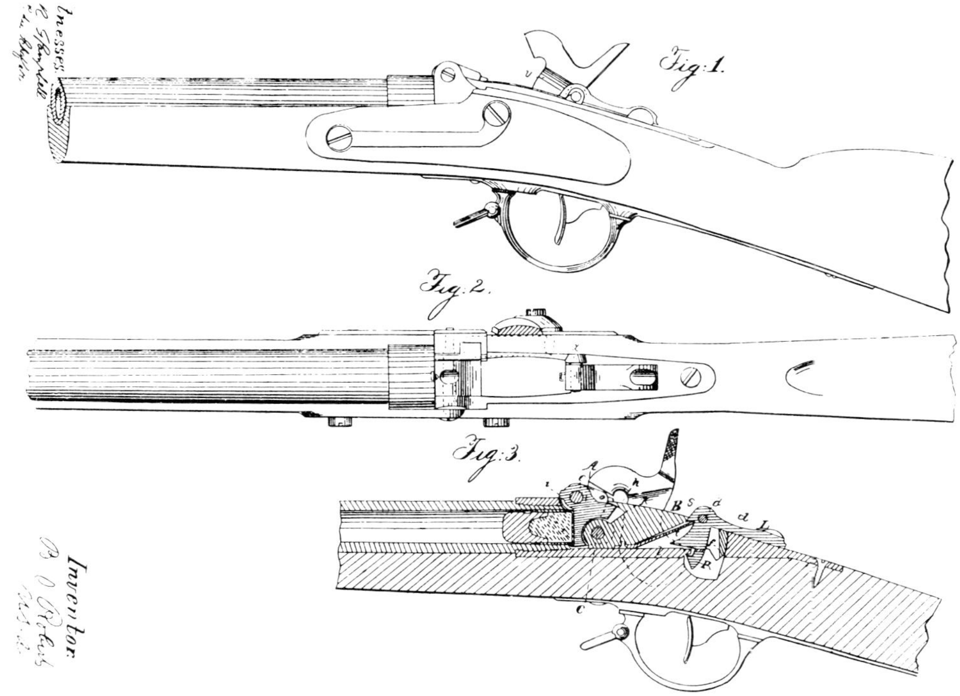

The Mont-Storm rimfire conversion was essentially the same in appearance as his original percussion model with front-hinged, tip up breechblock. The rimfire type, however, was made with a solid block and moveable firing pin within. The U.S. Government paid the inventor $2500 in September 1858 for the right to alter 2000 muskets using his action. The work was begun at Harpers Ferry Arsenal but terminated after a few arms were completed because the arsenal quickly fell to the Confederate forces at the opening of the Civil War. The demand for regular arms forced postponement of any further breech-loading modifications of existing muzzle-loaders.

The B.S. Roberts patent of February 27, 1866 employed a similar front-hinged tip over breechblock arrangement operated by lifting up on the lever lying along the tang. The copy of the original patent drawing is interesting because it reflects the actual shape of the cartridge used in this conversion of the .69 caliber muzzle-loading rifles. The third cartridge shown opposite was used in another experimental action but which one remains to be determined. (Illustration from the original patent)

RFR 35 – .58in (14.73mm) U.S. Allin,
Miller, or Springfield, 1866

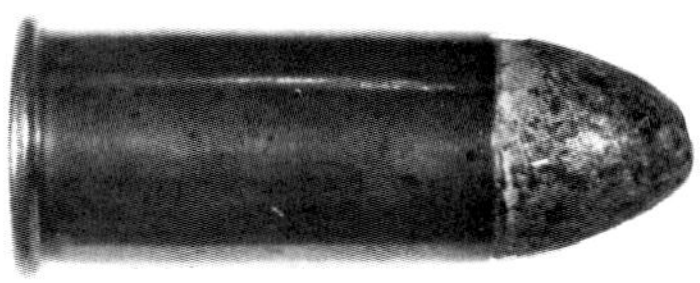

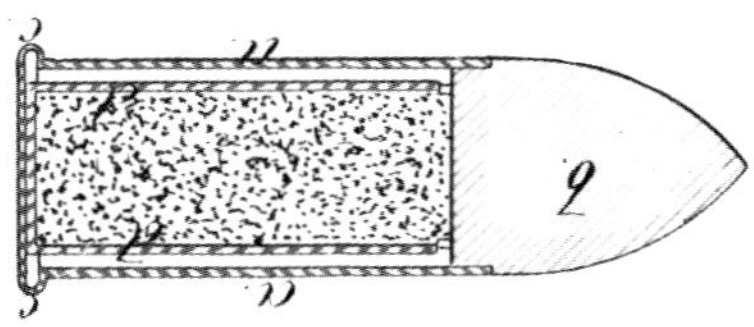

Description

(a) Copper 1.17in case, .589in lead bullet with no grooves showing, no h.s.

(b) Copper 1.20in case, .597in lead bullet, headstamp is "58 CAL." in two lines, denoting manufacture at Frankford Arsenal, Philadelphia, Pa.

(c) Same approximate case dimensions as above. C.J. Bergen's cartridge with double copper case made with channels between the two walls leading from the rim priming to the base of the bullet so that ignition of the powder occurred at the front of the charge. A .36in patent model had four channels, the .58in six channels. Made by Jacob Goldmark, New York, and may bear the "JG" headstamp.

(d) Copper 1.21in case, .597in lead bullet. Headstamp is "58 CAL." Cutaway specimen No. 120 at 1876 Philadelphia Exposition.

Comments

The .58 caliber Allin, Springfield, or Miller probably was used in the Chabot, Needham and other experimental transformed rifle-muskets tried out after the Civil War. The cartridge is often referred to as the .58 Musket. The slight case length variations which occur between the various makes of this cartridge do not warrant a conclusion that they are different cartridges. Unheadstamped specimens are relatively common in cartridge collections, but any of the headstamped rounds illustrated opposite are difficult to locate, and complete collections of them are few in number.

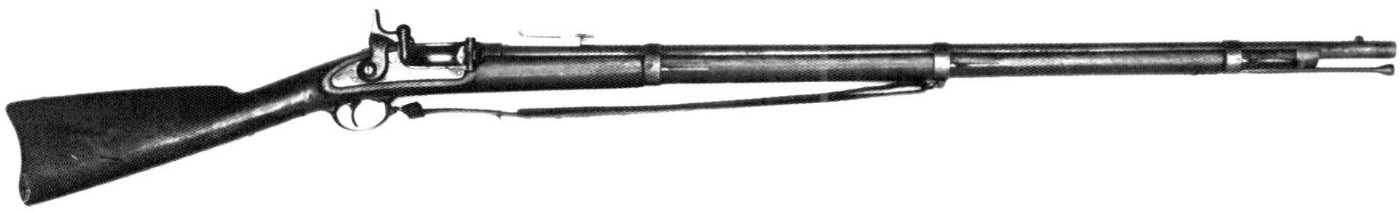

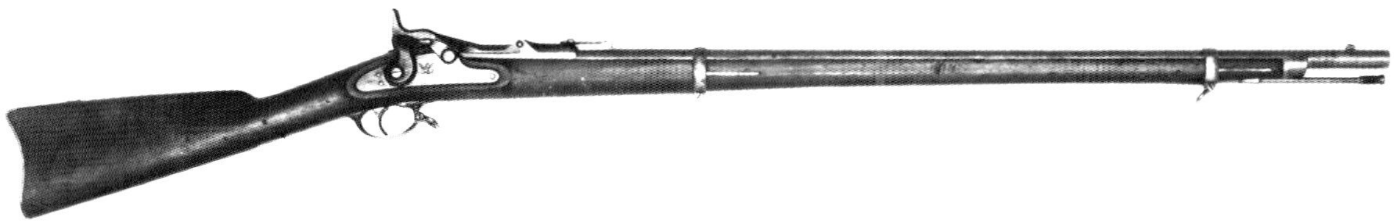

The breech-loading action of Joseph and George Needham patented in the United States on May 21, 1867, shown on the first rifle above, was also hinged at the front of the breechblock, but to the right side of the rear portion of the barrel. It swung out to the right to expose the chamber for loading. This solid assembly containing a rimfire firing pin was screwed onto the cut off breech end of the muzzle-loading barrel. It was one of the unsuccessful competitors of the action patented on September 19, 1865 by Erskine S. Allin, Master Armorer at Springfield Arsenal. The "trapdoor" system, shown second, bears striking resemblance to the earlier Mont-Storm action. Ultimately, Allin's action was adopted after trials of many systems and remained the basic breech-loading action used to manufacture the early .58 rimfire, as well as the .50 and .45 caliber rifles that followed in the United States service.

Shown below are the known headstamps applied to the .58 caliber rimfire ammunition used in the Allin converted rifles and most others entered in the trials. From left – C.D. Leet, Fitch Van Vechten, Jacob Goldmark, Springfield or Frankford Arsenals, Union Metallic Cartridge Company. Many were manufactured with no headstamp. (Rifle photos courtesy Milwaukee Public Museum)

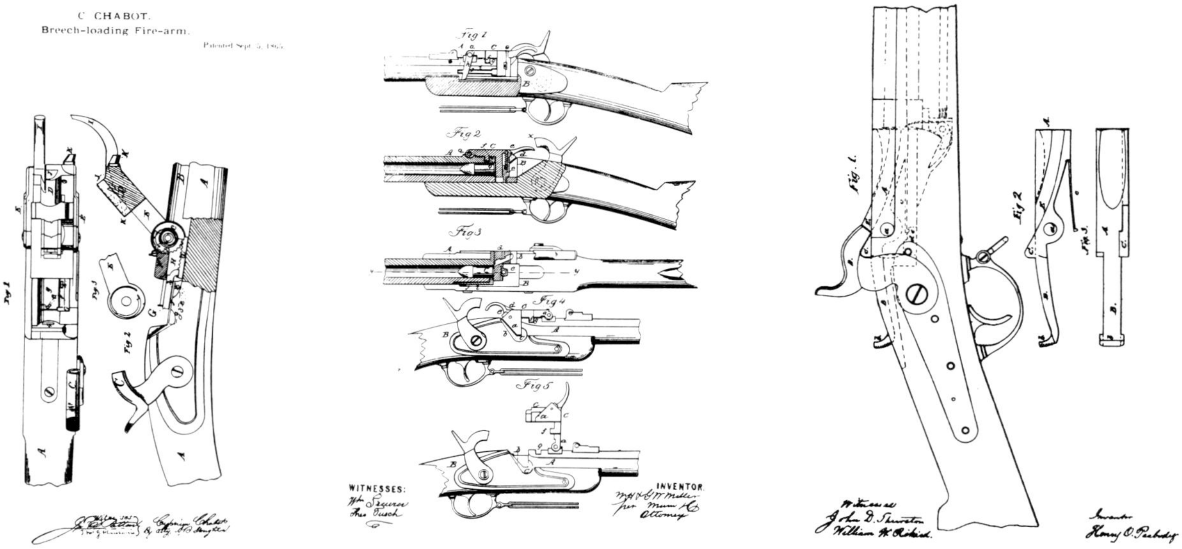

Shown here are three other actions entered in the trials leading to the selection of the Allin action for converting muzzle-loaders to rimfire breech-loaders. The Chabot, left, and the Miller, center, both tipped up for loading. The Peabody used the same rear-hinged, dropping down block action employed on his other rifles but the operating lever was fitted to the rear of the block. Lifting up on it lowered the block to expose the chamber. (From the original patent drawings)

RFR 36 – .58in (14.73mm) U.S. "Long Musket" for experimental rimfire conversion, ca. 1866

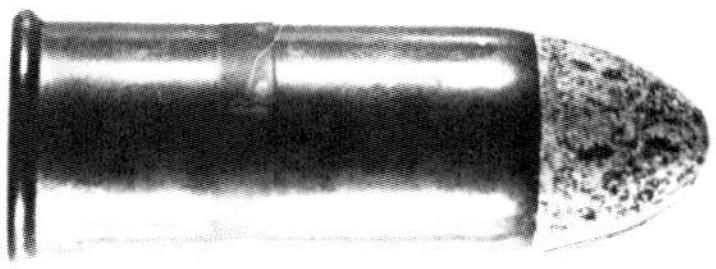

Description

Copper 1.389in case, .605in lead bullet, no h.s.

RFR 37 – .58in (14.73mm) U.S. Roberts for converted musket, 1866

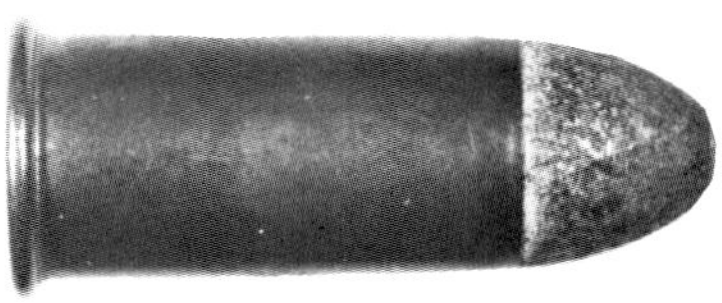

Copper 1.294in case, .602in lead bullet. Headstamp is a raised "H" in depressed circle.

Comments

The experimental "Long Musket" is distinguishable from the Roberts cartridge by its slightly longer case but considerably smaller rim, head, and case mouth dimensions (See dimensions chart, end of rimfire chapters). The Roberts was the largest of the cartridges for .58 caliber converted muskets in case dimensions, though the Mont-Storm was longer overall.

Shown below are a .58 caliber U.S. rifle-musket and a .577 caliber British Enfield rifle both converted to breech-loader with the Roberts action. Such rifles will be found made for the .58 Roberts centerfire cartridge of identical dimensions. (Milwaukee Public Museum photos)

RFR 38 – .50-70in (12.70mm) U.S. Springfield 1866 Modified Allin Conversion of the .58in Rifle-Musket

Description

(a) Copper 1.73in case, .521in two-groove pointed lead bullet. No h.s.

(b through g) Cutaway specimens from Drawer 6, U.S. Ordnance ammunition display at Philadelphia Exposition, 1876. Modification of charge of Hazard's FFG powder, 70 grains varying in the amount of solidification from .05 (top specimen) through .3 at bottom. As the powder charge was progressively compressed, the case length was reduced correspondingly to the length necessary to accommodate the powder, resulting in the following lengths: (b) 1.67in; (c) 1.622; (d) 1.567; (e) 1.515; (f) 1.468; (g) 1.422.

Solidification applied to the specimens was as follows: (b) .05; (c) .1; (d) .15; (e) .2; (f) .25; (g) .3.

Another set in the display employed FFFG powder but was otherwise identical. None of these compressed powder cartridges was adopted.

Comments

The .50-70 Springfield rimfire was made for those transformed .58in rifles with barrels lined to reduce the caliber. The cartridge had hardly been developed when U.S. military authorities turned to centerfire as a better alternative for self-contained ammunition. Therefore, the rimfire Springfield was shortlived. Some Sharps percussion rifles and carbines were converted to chamber the .50-70 rimfire, but these were probably few in number in view of the short life span of the cartridge.

RFR 39 – .44-45-350 (11.18mm) U.S. Experimental, 1865-1866

Description

Copper 1.136in case, .466in lead bullet, no h.s.

RFR 40 – .44-40-300 (11.18mm) U.S. Experimental, 1865-1866

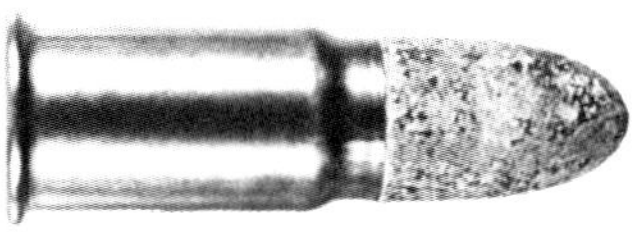

Copper .998in case, .454in lead bullet, no h.s.

RFR 41 – .44-40-350 (11.18mm) U.S. Experimental, 1865-1866

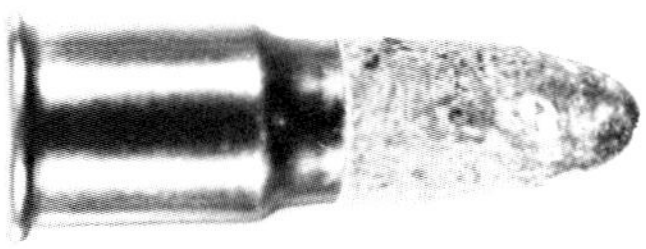

Copper .881in case, .460in lead bullet, no h.s.

RFR 42 – .44-40-300 (11.18mm) U.S. Experimental, 1865-1866

Copper .881in case, .461in lead bullet, no h.s.

Comments

These cartridges are a part of the experimental series of rimfire rifle loads developed at the National Armory (Springfield Arsenal) just following the Civil War. They appeared in the display prepared by U.S. Ordnance department for the Philadelphia Exposition of 1876 as numbers 126 through 129 in cutaway form. Only test barrels or rifles would have been made for them.

The designations reflect nominal bore diameter, powder charge, and weight of the bullet.

Reference: "Small Arms Ammunition at The International Exposition, Philadelphia, 1876." Compiled in 1972 by Col. Berkeley R. Lewis and published by the Smithsonian Institution.

RFR 43 – .44-45-500 (11.18mm) U.S. Experimental, 1865-1866

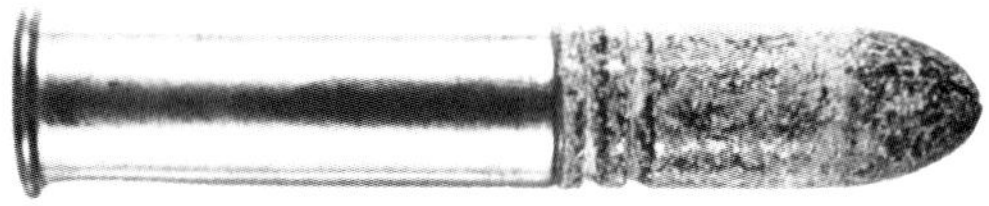

Description

Copper 1.409in case, .459in lead bullet, no h.s.

RFR 44 – .44-50 (11.18mm) U.S. Experimental, 1865-1866

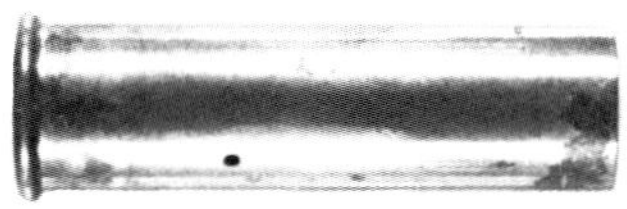

Copper 1.559in case, unprimed, with hole in side of case. No h.s.

RFR 45 – .44-55-500 (11.18mm) U.S. Experimental, 1865-1866

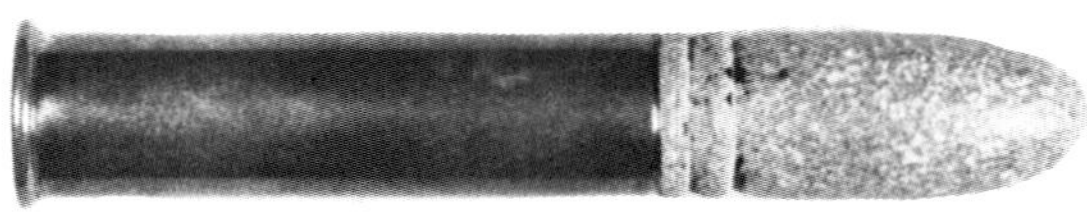

Copper 1.675in case, .460in lead bullet, no h.s.

RFR 46 – .44in (11.18mm) U.S. Unknown, ca. 1865-1866

Copper 1.025in case, .469in lead bullet, no h.s.

RFR 47 – .44in (11.18mm) U.S. Unknown, ca. 1865-1866

Copper 1.025in case, .450in lead bullet, no h.s.

Comments

The first three cartridges above are part of the same series shown on the previous page. The upper specimen was No. 125 in the display at the 1876 Philadelphia Exposition. The fourth and fifth specimens are not known to be military experimentals and conceivably could have been proto-types of the .56-46 Spencer sporting rifle cartridge developed after the Civil War. (See examples of sporting cartridges developed after military rounds, end of rimfire section).

Description

RFR 48 – .46in (11.68mm) U.S. Experimental, ca. 1865-1866

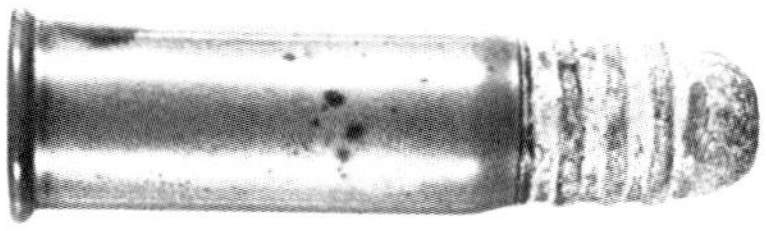

Copper 1.327in case, .469in lead bullet, no h.s.

RFR 49 – .50in (12.7mm) U.S. Experimental, ca. 1865-1866

Copper 1.059in case, .518in lead bullet, no h.s.

RFR 50 – .50-50-500 (12.7mm) U.S. Experimental, 1865-1866

Copper 1.135in case, .512in lead bullet, no h.s.

RFR 51 – .50-55-500 (12.7mm) U.S. Experimental, ca. 1865-1866

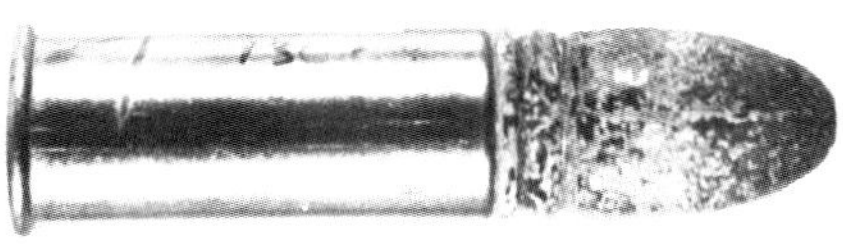

Copper 1.247in case, .516in lead bullet, no h.s.

Comments

Charles Suydam illustrates the upper specimen in "The American Cartridge," plus another with the same case but longer bullet. It and the second cartridge shown are believed to be Frankford Arsenal experimentals. The third and fourth are known to be of that origin. The .50-55-500 is specimen No. 130 in the Philadelphia 1876 Exposition display.

RFR 52 – .50in (12.70mm) U.S. Experimental, ca. 1866(?)

Description

Copper bottle-necked 1.381in case, .498in lead bullet, no h.s.

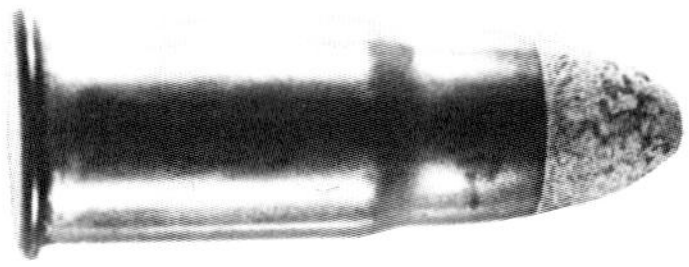

RFR 53 – .50-65-480 (12.7mm) U.S. Experimental, 1865-1866

Copper 1.336in sectioned case, .515in lead bullet. Headstamp is "50 CAL."

RFR 54 – .41-60-312 (10.4mm) Swiss, 1866-1867

Copper 1.492in sectioned case, .425in lead bullet. Headstamp is an impressed "T."

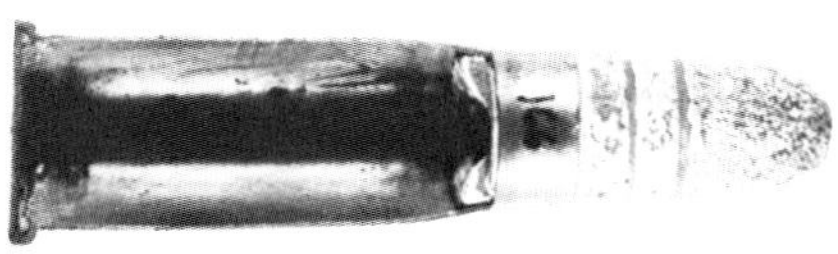

RFR 55 – .65in inside diameter of case mouth (16.51mm) Unknown

Copper 2.28in primed case.

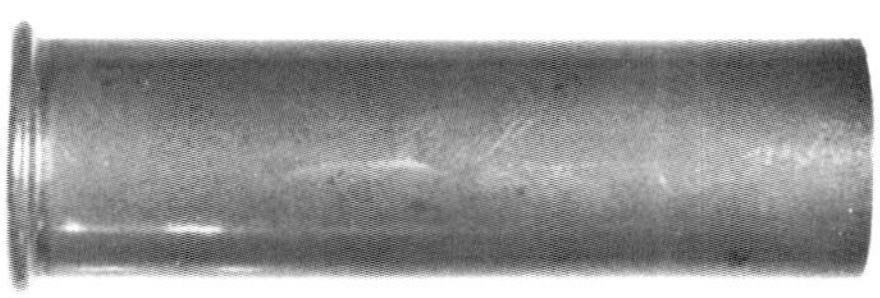

Comments

The upper cartridge was identified at one time on a dealer's list as "NY Trials." However, this is not confirmed and no other information has been unearthed about it. It resembles the .50 Jenks carbine shown previously but the case is somewhat longer, the shoulder is positioned differently than the Jenks, and the crimp extends well over the largest part of the bullet. The second specimen is No. 122 in the Philadelphia 1876 Exposition as it was shown sectioned. The third is of Swiss manufacture and shown because it was displayed as No. 166 at the exposition. The last specimen is a primed 20 gauge rimfire case and speculation is that it may have been intended for use in a military musket rebarreled for forager use. Military-marked shot cartridges were later made by Frankford Arsenal in similar dimensions for rebarreled Springfield rifles.

RFR 56 – .50in (12.70mm) U.S. Crispin for Modified Smith Carbine, 1867

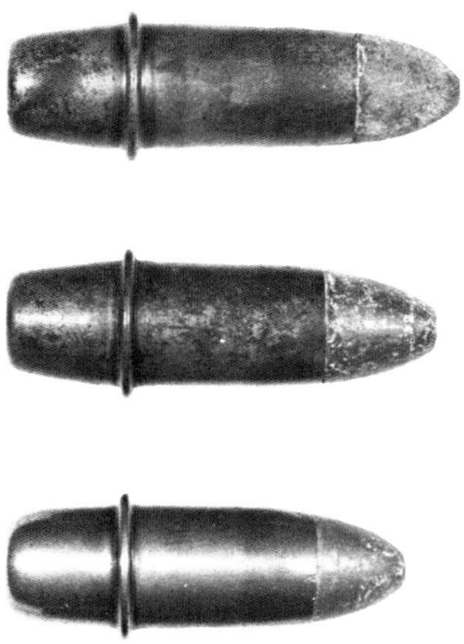

Description

(a) Copper 1.50in case, .665in annular rim around case body in which the priming composition is located, .520in bullet, no headstamp or other identifying marks.

(b) Same construction as above but 1.38in case length.

(c) Same construcion as above but 1.34in case length.

Comments

Although it is peculiar in appearance the Crispin is properly classified as a rimfire. The rim which contains the fulminate is located around the body of the case. The drawing from Silas Crispin's 1867 patent modifying the percussion Smith carbine to use these cartridges shows that the rear portion of the case fits in the breech, the front part in the chamber. Specimens have been examined with a shorter base section than those above but it is evident they would not fill the breech end of the chamber for which those illustrated above were made. Reproductions of Crispin cartridges exist and any purchaser should demand a written guarantee of authenticity. Pistol sizes exist also in .44, .36, and .31in.

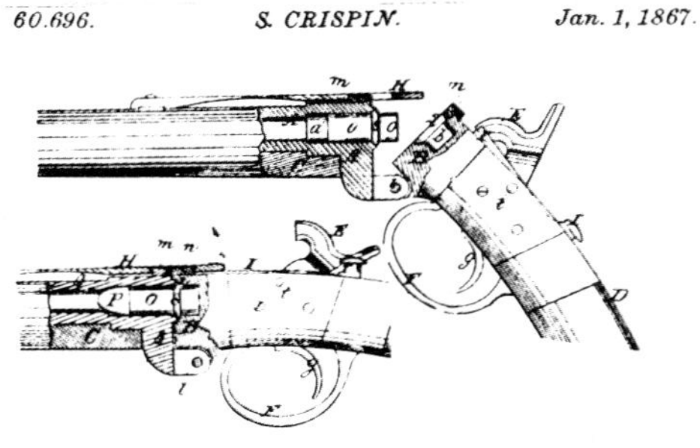

Description

RFR 57 – .52in (13.21mm) Unknown

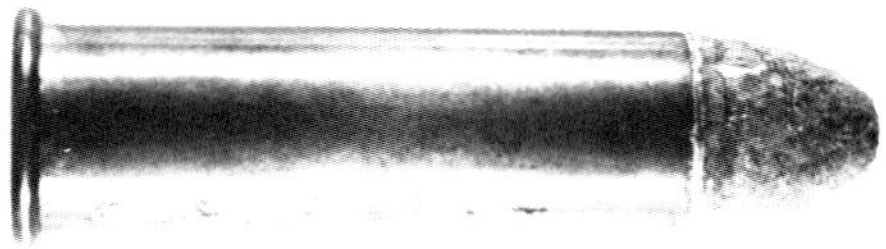

Copper 1.786in case, .538in lead bullet, no h.s. Similar to the.50-70 rimfire but larger overall.

RFR 58 – .50in (12.7mm) U.S. Unknown

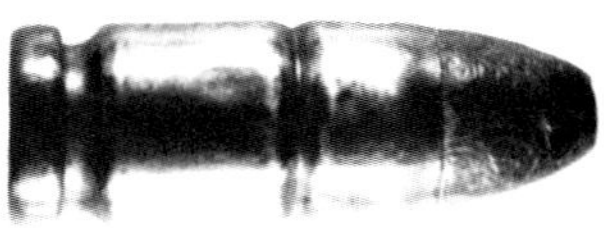

Copper rimless 1.131 case crimped over largest diameter of a .514in lead bullet, deep cannelure below the bullet. Head has a depression around a central raised portion. No h.s. May be a Burton experimental.

RFR 59 – .44in (11.2mm) U.S. Winchester Experimental, Dupee patent

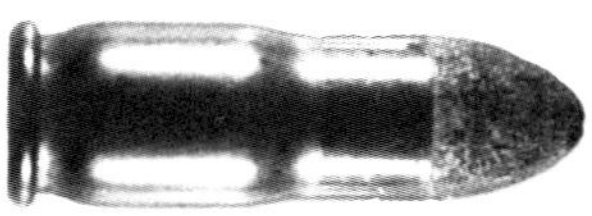

Copper 1.089in rimless, necked case, .454 lead bullet,no h.s. A companion .44 Dupee rimless rimfire with slightly longer case also exists.

RFR 60 – .56-56 (13.21mm) U.S. Spencer, Winchester commercial

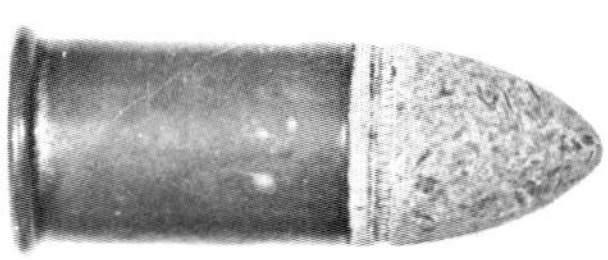

Copper .872in case, .535in single-groove lead bullet. Metric dimension reflects the bore diameter of the rifle. Headstamp is a raised "H" in impressed circle.

RFR 61 – .56-52 (12.7mm) U.S. Spencer, Winchester commercial shot load

Copper 1.036in case, .490in straw-colored paper bullet containing bird-shot. Headstamp is the raised "H" in depressed circle. Usable in .56-56 and.56-50 arms.

RFR 62 – .56-50 (12.7mm) U.S. Spencer, Winchester shot load

Description

Copper 1.544in extended case containing shot load covered by a cardboard wad. Headstamped is impressed "H."

RFR 63 – .56-46 (11.18mm) U.S. Spencer, Fitch Van Vechten commercial

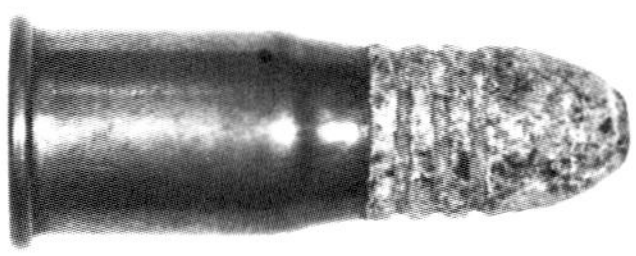

Copper .910in case, .474in lead bullet with three grooves. Early short-necked version with "F.V.V.&Co." h.s.

RFR 64 – .56-46 (11.18mm) U.S. Spencer, Winchester commercial load

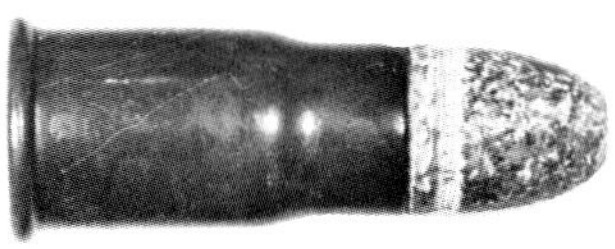

Copper 1.032in case, .466in single-groove lead bullet. Headstamp is a raised "H" in depressed circle.

RFR 65 – .46in (11.68in) U.S. Extra Long

Copper 1.593in case, .461in lead, two-groove bullet. No h.s., probably made by Union Metallic Cartridge Company.

RFR 66 – .46in (11.68mm) U.S. Long, Commercial Load of .46 Remington Carbine

Copper 1.25in case, .454in single-groove lead bullet. Headstamp is raised "H" in impressed circle.

Comments

RFR 64 is typical of the .56-46 Spencer carbine commercial loads made after the Civil War. Others similar with different case lengths exist, still unidentified. The metric figures following the Spencer cartridge identities represent the nominal bore diameter.

Description

RFR 67 – .44in (11.18mm) U.S. Extra Long

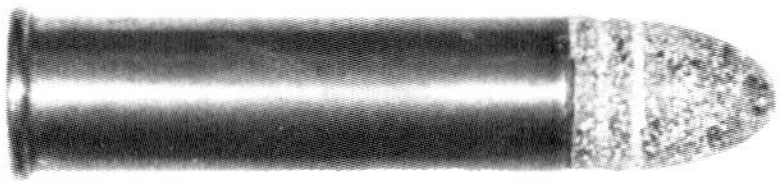

Copper 1.50in case, .439in lead, two-groove bullet. Raised "U" in impressed circle (Union Metallic Cartridge Company).

RFR 68 – .44in (11.18mm) U.S. Ballard

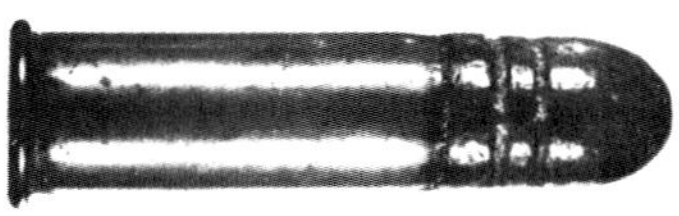

Copper 1.13in case, .444in lead, two-groove bullet. No h.s., but made by U.S. Cartridge Co.

RFR 69 – .44in (11.18mm) U.S. Long

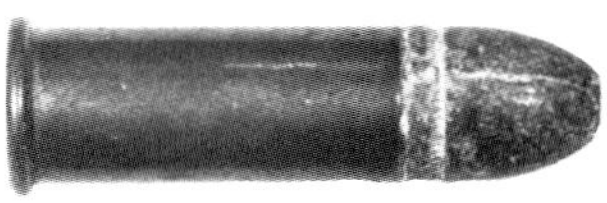

Copper .932in case, .440in lead, single-groove bullet. Raised "H" in impressed circle, by Winchester Repeating Arms Co.

RFR 70 – .38in (9.65mm) U.S. Extra Long

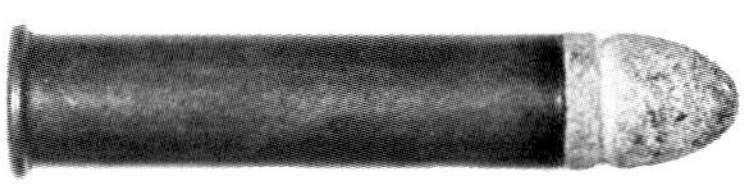

Copper 1.490in case, .379in lead bullet. Raised "H" in impressed circle.

RFR 71 – .32in (8.13mm) U.S. Extra Long

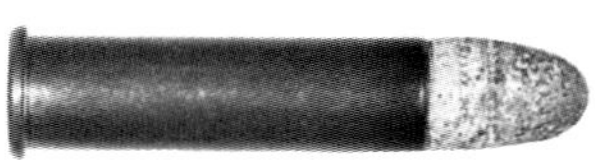

Copper 1.12in case, .318in lead, single-grooved bullet. Raised "H" in impressed circle.

Comments

The commercial cartridges on this and the previous two pages are only a sample of those that exist mady by a number of companies in the United States and other countries. The reader is referred to "The American Cartridge," by Charles Suydam and the collector cartridge lists in the "Gun Report Magazine" every other issue. These sources provide information on commercial rimfire cartridges not covered here, and the latter is a guide to current collector values.

12 THE MILITARY RIMFIRE IN EUROPE

The large caliber rimfire cartridge was soon replaced by the centerfire in the United States after the Civil War. However, several European countries found the rimfire entirely suitable for transforming their huge stocks of obsolete muzzle-loading rifles to breech-loaders. The Swiss Milbank-Amsler and Danish Snider cartridges with their heavy, low-velocity bullets were only a stop-gap measure at best, but the Swiss, Norwegians, Swedes, and Danes all found improved, higher-velocity rimfire loads appropriate for their regular issue infantry rifles. Along with the American-developed cartridge, they adopted American weapons systems. The tubular magazine on the Swiss Vetterli was taken from the Henry and Model 1866 Winchester rifles and the Remington rolling block adopted by the Scandinavians was of American origin, made in those countries under Remington license.

European rimfires have received little publicity in the United States, probably because specimens have not been available to attract the attention of collectors. However, they represent a substantial part of the rimfire field, and are just as interesting as the American-made specimens. In fact, the 10.4x38R Swiss offers more loading, case, and headstamp variations than any other rimfire cartridge.

This section identifies the specimens primarily in the metric system, the first number identifying the bore diameter of the rifle and the second the case length. Bullet diameters vary from the bore diameter since most bullets were somewhat larger than the bore so that a proper gas seal would occur and the bullet take the rifling.

References: "Eidgenössische Handfeuerwaffen," Hugo Schneider and Michael am Rhyn, 1979
"Handfeuerwaffen System Vetterli," Hugo Schneider, Michael am Rhyn, Oskar Krebs, Christian Reinhart, Robert Schiess, 1970
"Cartridges for Collectors," Vol. III, Fred A. Datig, 1967
Correspondence and research records of Messrs. George Kass and Victor v.B. Engel

RFR 72 – 18mm (.709in) Swiss Milbank-Amsler, 1867 Conversion of Prelaz-Burnand rifle

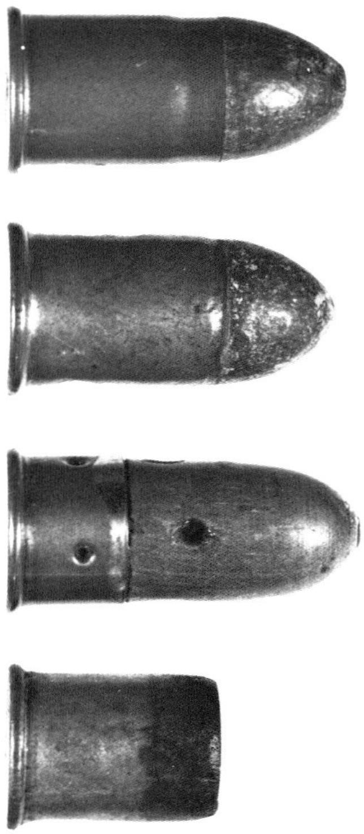

Description

(a) Copper 25mm case, 17.86mm lead bullet with single groove covered by the case, which is slightly necked by the crimp. Headstamp is a triangular depression near the edge of the head with a raised dot in its center.

(b) Copper 25mm case, 17.93mm bullet, no visible necking since case mouth is crimped into bullet's groove. Headstamp is an off-center depressed Swiss cross with a raised "B" within.

(c) Copper 13.9mm case, 17.68mm wood bullet dummy for above rounds. Headstamp is an impressed triangle with one dot in its center.

(d) Copper 24mm case, blank charge for above with white card wad over powder. Same headstamp as (c) above.

Comments

The 18mm Milbank-Amsler cartridge is the largest adopted by any country for conversion of muzzle-loading rifles to rimfire. The cartridge also exists in brass case ball load and extended brass case shot with raised "ST" headstamp made by Stahel of Switzerland who made commercial ammunition in many calibers.

These and the Swiss cartridges illustrated on the following pages are found with from one to four dots within or on either side of the headstamp. The dots indicated the number of the loading machine.

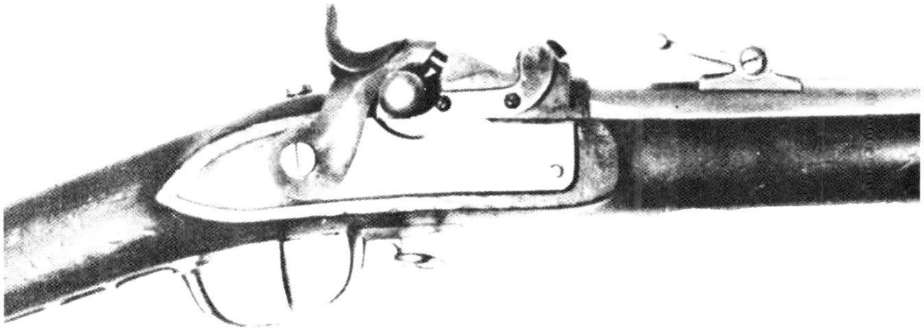

In 1867, the Swiss adopted the breech-loading system invented by the American Isaac M. Milbank and evidently perfected by Jakob Amsler, a Swiss, for acceptance by his country's military authorities for conversion of existing muzzle-loading rifles to rimfire breechloaders. The version of this Milbank-Amsler action illustrated above was applied to the 18mm Prelaz-Burnand rifle, Model 1842/59, the original cartridge for which is shown on page 42.

The Milbank-Amsler was a hinged breechblock system tipping up and forward. In order to convert the muzzle-loader, the rear part of the barrel was cut off and threaded to receive this new breechblock. The block housed a rim firing pin which struck the cartridge rim on the right side. The left side of the block incorporated a hook extractor which levered the empty case out of the chamber on opening.

These large-bore rimfire rifles, as in other countries, were a stopgap measure to make some use of the large number of such muzzle-loading rifles in service at the time self-contained ammunition was adopted. However, Schneider and Am Rhyn state that 56,369 were ultimately converted.

Some of these arms probably originated as flintlocks, which were converted to percussion ignition around 1842. Then they were converted from smoothbore muskets to rifles commencing in 1859, and finally transformed from muzzle-loaders to rimfire breech-loading rifles in 1867.

RFR 73 – 10.4x38R (.409in) Swiss for Milbank-Amsler Conversion of the Federal Rifle, Mod 1867

Description

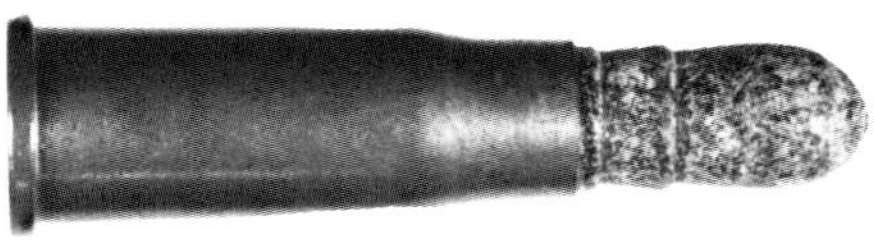

(a) Copper case, 10.18mm muzzle-loading style 1863 bullet with shallow grooves, no headstamp. This cartridge appears to be an original factory load and may be first type tried in the conversion.

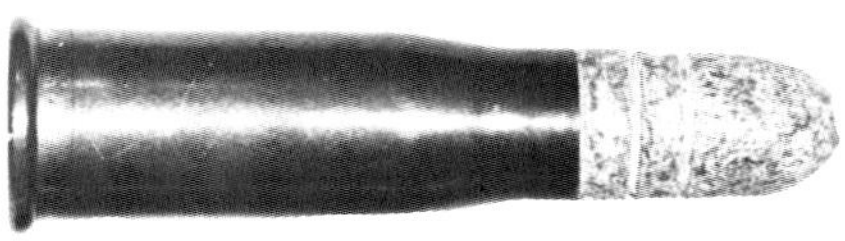

(b) Copper case, 10.62mm two-grooved lead bullet, headstamp is large indented cross. This is the 1867 cartridge for the converted muzzle-loader, but would also chamber in the later Vetterli rifle.

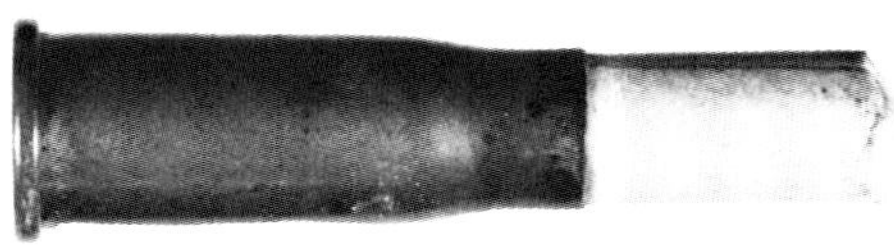

(c) Copper case, 10.72mm paper-covered column of five hollow-based slugs nested together. A guard load. Headstamp is "T" with two dots.

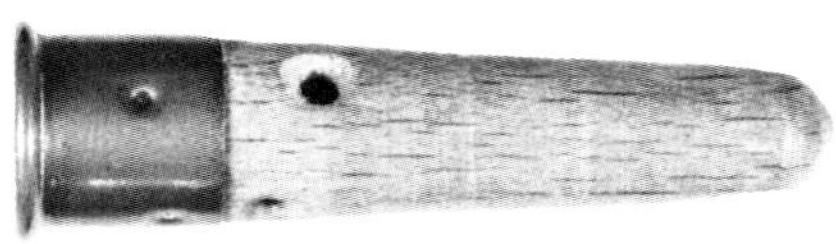

(d) Copper 13.72mm case, extended wood bullet. A functioning dummy which probably would also chamber in the Vetterli bolt-action rifle. Headstamp is an impressed cross.

Comments

The earliest 10.4x38R Swiss cartridges for the converted Federal rifle were loaded with grooved, naked lead bullets, not the paper-patched type to come later. Specimen (a) above suggests that the first cartridges tried in this Milbank-Amsler breech-loader retained the smaller-than-bore size bullet used in the muzzle-loader. Evidently, it did not take long for the Swiss to realize that a larger-than-bore bullet should be used which would fill the grooves as it was forced into the rifling.

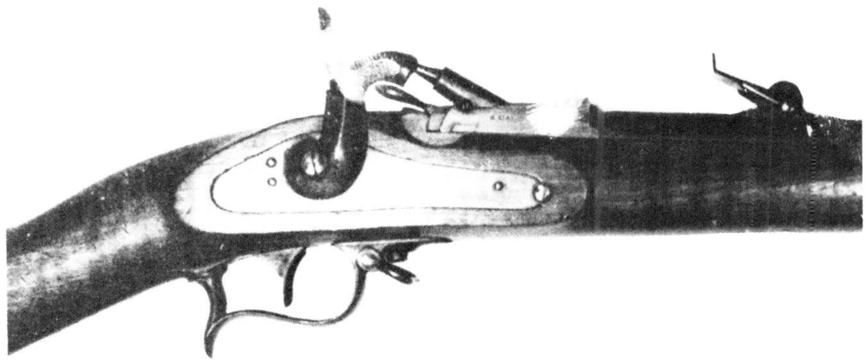

The Model 1863 muzzle-loading Swiss Federal rifle of 10.4mm formed the basis for this conversion to rimfire cartridge. The same action as that used in the larger-caliber conversions was applied with substantial modifications. The round knob operating lever was replaced with a more gracefully contoured thumb lift type, still situated on the right below the hammer. The firing pin housing sits up considerably higher than the original form of the breechblock, and the hinge arrangement has been altered to do away with the arm on the original action held to the right side of the block by two screws.

Messrs Schneider and Am Rhyn state that 76,676 of the 10.4mm muzzle-loading rifles in all forms were eventually transformed to breech-loading rimfire.

RFR 74 – 10.4x38R (.419in) Swiss for Vetterli Rifle, Models 1871-1878

Description

(a) Copper case, 10.95mm (over paper patch) lead bullet, headstamp is the Swiss cross in an impressed circle. This is the Vetterli rifle load.

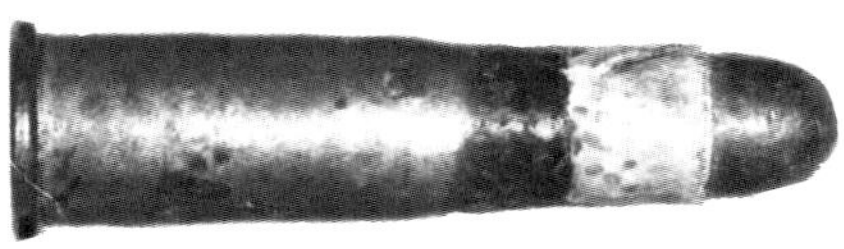

(b) Same case as above. Paper patch does not extend over neck of the case. Case neck is lubricated with grease. This load was used in the Vetterli carbine.

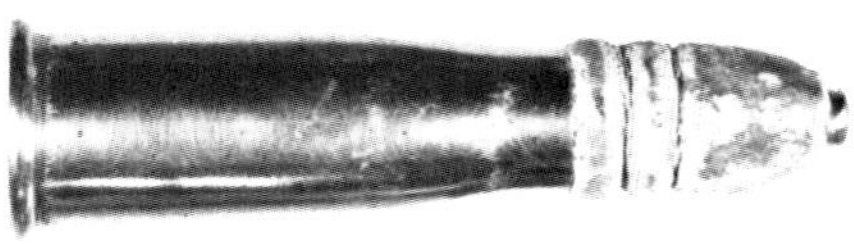

(c) Same case as above loaded with a two-groove bullet containing a copper tube filled with explosive compound intended to detonate on impact. Swiss cross headstamp and an embossed "G" on the nose of the copper tube suggesting that this component was manufactured by the French ammunition maker, Gevelot.

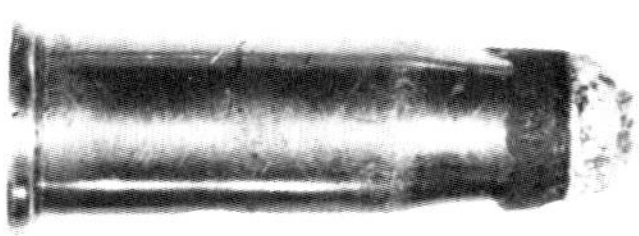

(d) Same case as above loaded with round ball, no h.s. Presumed to be a gallery practice cartridge.

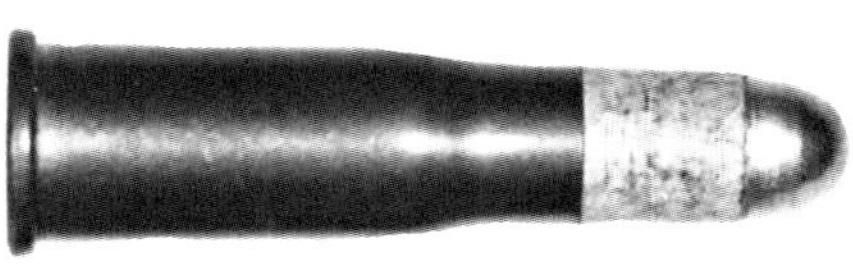

(e) Same case as above loaded with an experimental armor piercing steel, round-nosed, paper-patched bullet. A pointed bullet version also exists. Cross in a circle.

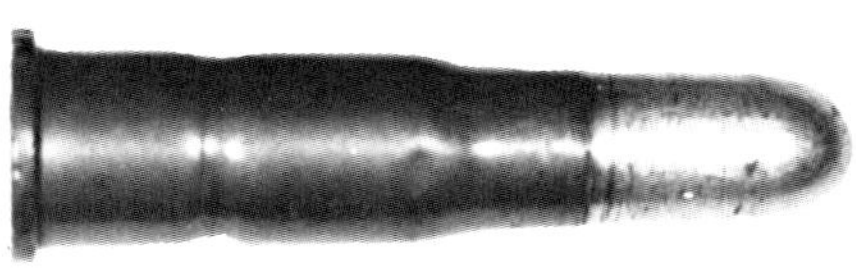

(f) Same case as others above but constructed with a heavy steel bullet soldered into the case mouth flush with the neck, and a deep cannelure around the case body. This is an inert exercise dummy.

RFR 74 – 10.4x38R Swiss Vetterli

Description

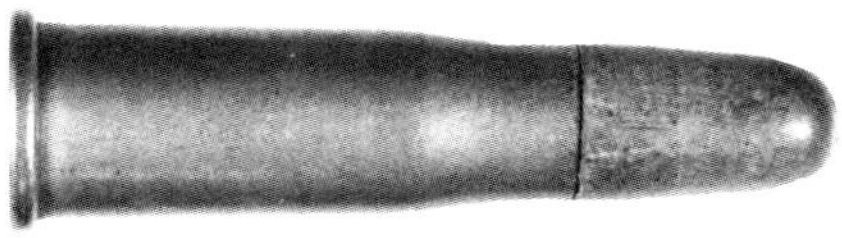

(g) Copper case, 10.92mm wood bullet headstamp "T" (for Thun arsenal). A blank loading.

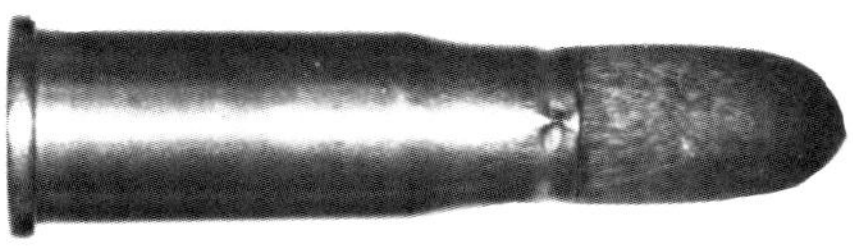

(h) Same as above, bullet dyed purple and case heavily stab crimped into mock bullet. Headstamp is the Swiss cross with a raised dot in the end of each arm of the emblem indicating the loading machine on which the cartridge was manufactured.

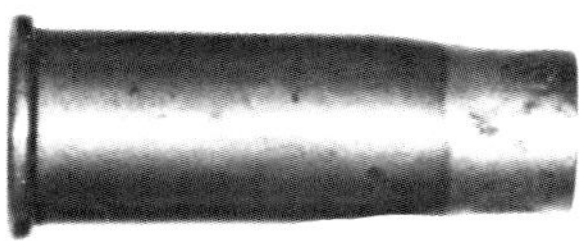

(i) Case as above closed over a blank charge with a shellacked wad of linen. Headstamp is the cross with two dots. Intended for single loading only.

(j) Case as above with heavy grease over sharply crimped in neck. The blank charge is closed with a papier mache wad, h.s. is the very early large Swiss cross indicating this blank was loaded ca. 1867. Suitable only for single loading.

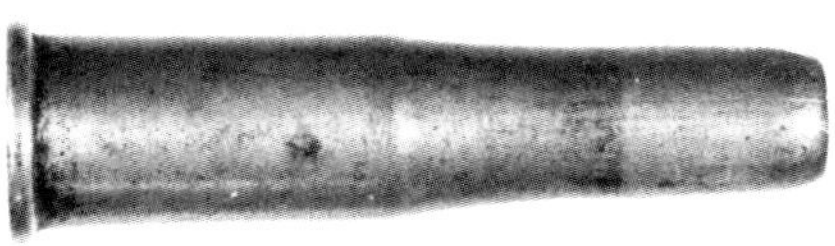

(k) Brass case drawn to the full length of a loaded cartridge and closed with crimp over a cardboard wad. This blank was evidently intended to be worked through the tubular magazine of the Vetterli rifle. Headstamp is an impressed cross within a circle.

Comments

The 10.4x38R Swiss Vetterli cartridge was in service from 1867 until replaced by a 7.5mm centerfire cartridge used in the Schmidt-Rubin rifle which appeared in 1889. Consequently, more standard and experimental loadings were developed for it than any other rimfire rifle round. Most European and American ammunition companies manufactured it and a large variety of headstamps and commercial loadings exist for the collector.

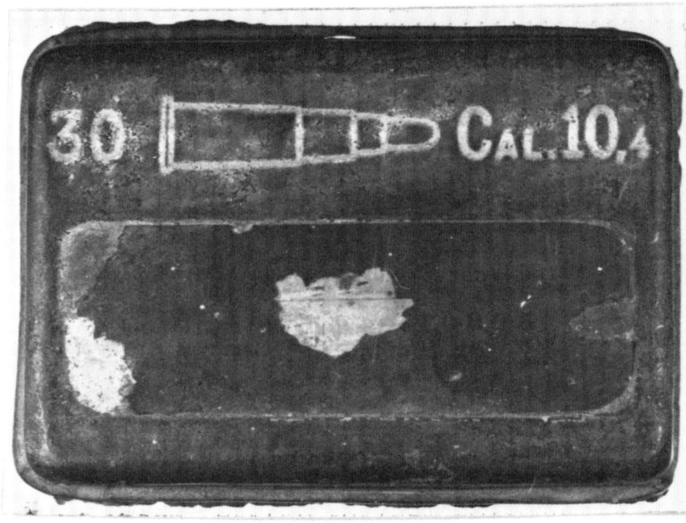

The tin container above, soldered closed against moisture for field issue, contained three of the ten-round packets of 10.4x38R Swiss rimfire military cartridges. Packet labels printed in German and French indicate 10 metallic cartridges, small caliber, the one below packed September 7, 1874.

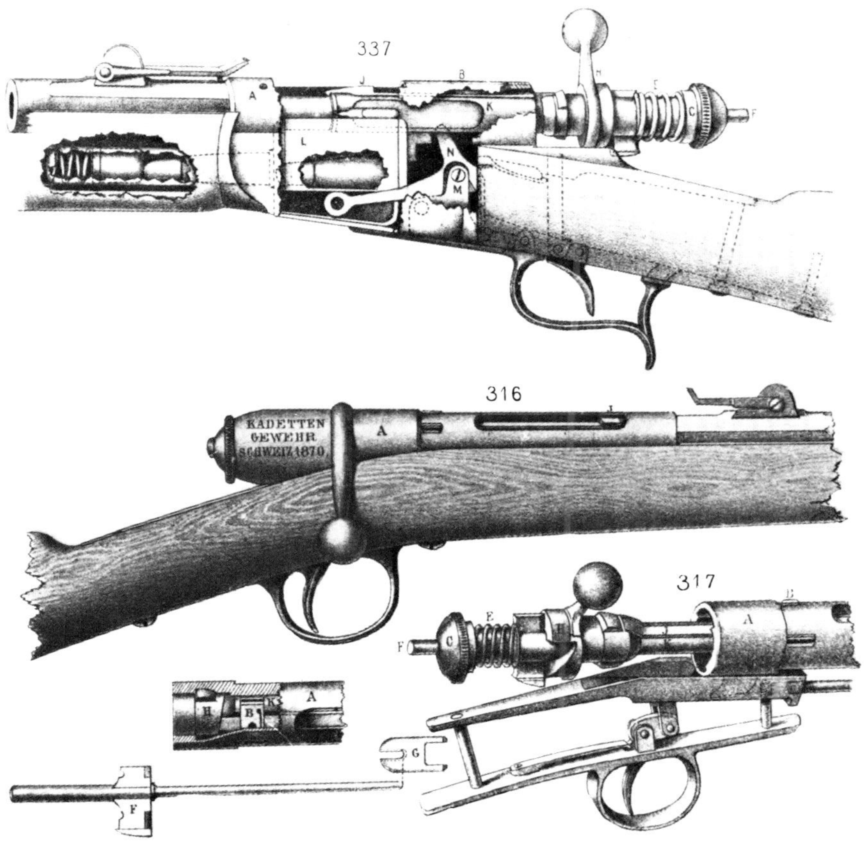

The earliest models of the Swiss Vetterli made in 1867 were made with an exposed hammer. This feature, and the tubular magazine, were applications of the same features found in the American Henry rifle. The Swiss, however, adopted the operating handle on the right rear side of the bolt rather than the underlever. The 1870 model dispensed with the separate exposed hammer and incorporated a spring loaded firing pin cocked by the operation of the bolt. Later models, including the Cadet rifle illustrated above, differed only in minor details. Loading 11 cartridges in the full-length rifle was accomplished in most models through a port in the right side of the action frame. With another cartridge in the carrier, and one in the chamber, the total capacity could be 13. The Swiss used this rimfire rifle until 1889-1890 when the 7.5mm Schmidt Rubin straight-pull bolt action rifle replaced it. (Drawings from Rodolphe Schmidt's "Les Armes A Feu Portatives," 1877)

RFR 75 – 17x32R (.669in) Norwegian Lund, Model 1867

Description

Copper case, 17.15mm conical lead bullet with greatest diameter covered by crimped over case mouth. No h.s.

RFR 76 – 12x28.6R (.472in) Norwegian Lund Carbine, Model 1869

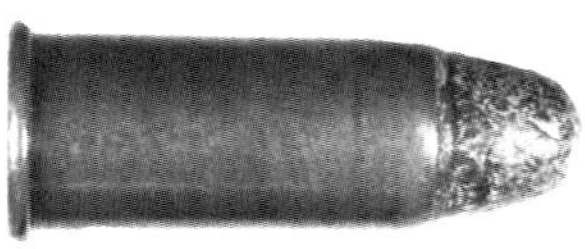

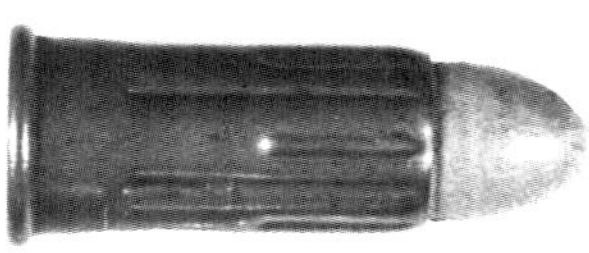

(a) Copper case, 12mm round-nosed lead bullet, no h.s.

(b) Copper case with six short and six longitudinal grooves, loaded with a red wood bullet, no h.s. This is a functioning dummy.

Comments

The Lund rimfires are among the rarest of cartridges in the military field. Norwegian Government records reflect that the 17mm was made in blank closed with a cork wad and a dummy. No description was available. A blank also was made for the 12mm carbine, closed with cork wad. Further experiments were conducted in 1876 with 10.5mm and 11mm cartridges for the carbine. These may have been necked down from the 12mm case type. However, no drawings or specimens of these are known to exist.

References: "Norwegian Small Arms Cartridges, Before 6.5mm Model 1894," Vidar Andresen, The International Cartridge Collector, Vol. Two, No. 4, June 1971.

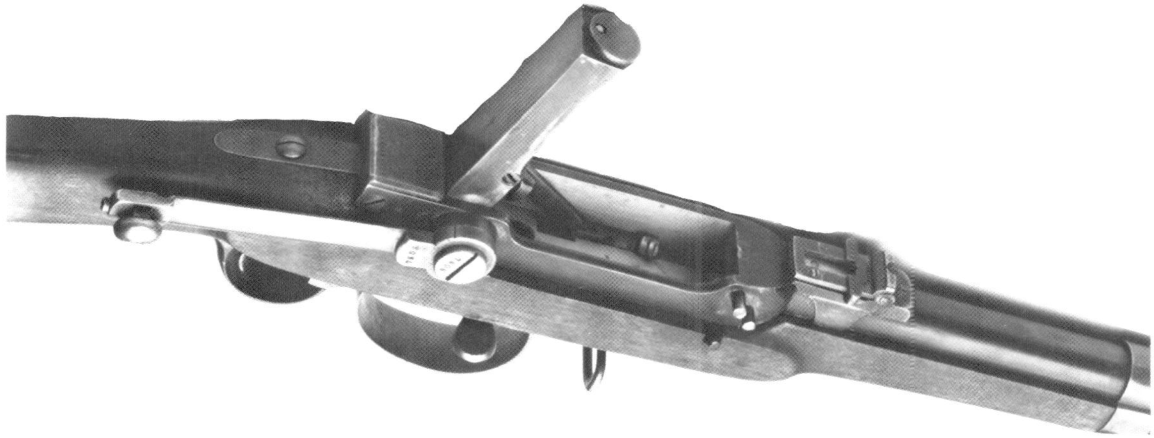

When rimfire ammunition had been perfected, Norway decided to convert its Model 1842 underhammer percussion breech-loaders (See PBL 3, p.57) to self-contained ammunition. Two systems of conversion were used. The primary system developed by Lund was accomplished by replacing the hollow percussion chamber with a solid breechblock containing a rim firing pin and working an extractor by a lever arrangement. The chamber for the metallic rimfire cartridge was bored out in the original barrel. Another system, Landmark's, was a bit different. He cut the original chambers in two, hinged them, cut a new chamber in the front part and put a firing pin and bolt face in the rear solid portion. Externally, the rimfire Lund and Landmark conversions appear little different from the original percussion breech-loading models. An obvious difference is the location of the rear sight on the rear portion of the Lund's barrel. It was positioned on the rear of the percussion rifle's breech.

The Model 1860/67 Lund was an issue infantry weapon in 17mm, and the 12mm carbine was issued to artillery and cavalry units. Several models of the 12mm carbine were made between 1865 and 1877. The Landmark conversion was made available only to civilian shooters, was never officially used by the military. (Research courtesy of Vidar Andresen, Nittedal, Norway. Photos from the Tøjhusmuseet–National Museum–Copenhagen, Denmark)

RFR 77 – 17.5x28R (.689in) Danish Snider, Model 1865

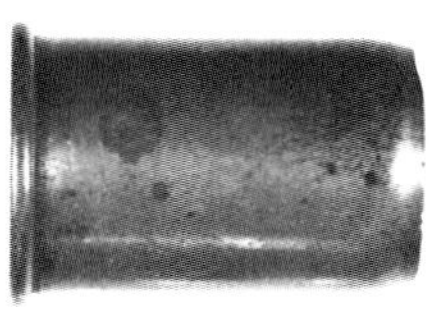

Description

(a) Copper case, 17.65mm bullet with straight ogive leading to a flat nose with case mouth crimped well over the largest diameter of the projectile. Headstamp is an impressed "T" for the Tojhus arsenal, Copenhagen.

(b) Brass base with a black cardboard case body loaded with wood bullet. This is the dummy round for the first type of the Danish Snider rimfire, which was soon replaced by the drawn copper type at (a). No h.s.

(c) Blank load for (a) closed with a cork or cardboard wad held in by case mouth crimp. "T" headstamp.

Comments

The Danish rimfire Snider cartridge has been described by other sources as having a 15.87mm bullet, undoubtedly because it was measured at the case mouth. This gives an incorrect reading of the bullet's true maximum diameter. When the muzzle-loading rifles which used these cartridges were transformed to the breech-loading system used, it was a simple matter to bore out the chamber to accept a case and bullet larger than the bore, thus ensuring that the bullet sealed the bore and properly took the rifling.

In 1865, Denmark adopted the breech-loading action invented by Jacob Snider of New York for converting its existing 17.5mm muzzle-loading pillar breech rifles, Model 1848, to rimfire breech-loaders. This is essentially the same action selected by Great Britain on September 18, 1866 for converting the Enfield rifle to centerfire breech-loader. Snider did not take out his American patent on this action until October 1867.

To operate this action, the operating knob on the left side of the breechblock was lifted to unlatch and turn the action open on its right-hand situated hinge. If a spent cartridge case was in the chamber, drawing back on the breechblock by the operating knob moved it to the rear on the spring-loaded hinge and drew back an extractor which started the empty case to a position where it could be removed or tipped out of the block recess. A fresh cartridge could then be inserted and the breechblock swung counterclockwise to close it for firing.

The primary difference between the Danish and British Snider is the angle at which the firing pin penetrates the block, the Danish being at less angle to strike the rim of the cartridge case. (Photos courtesy the Tøjhusmuseet–National Museum–Copenhagen, Denmark)

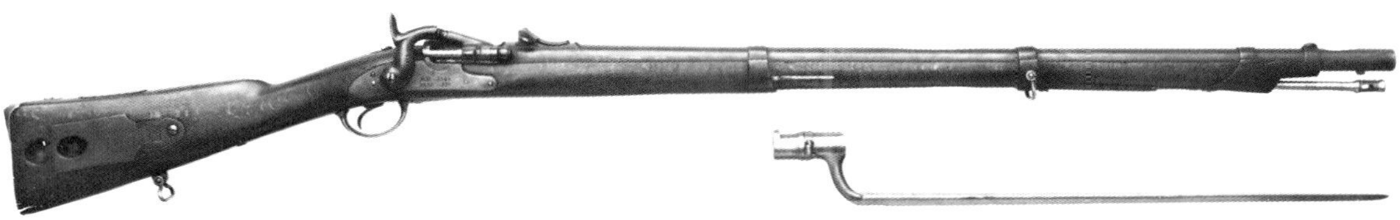

RFR 78 – 12.17x44R (.479in) Norwegian
Remington, Model 1871

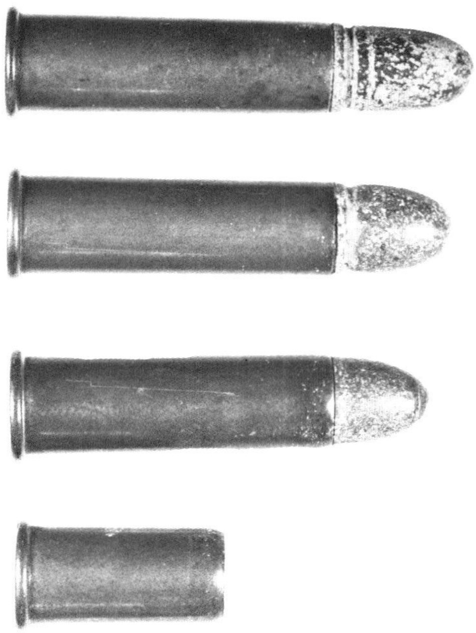

Description

(a) Copper case, 12.9mm single-groove round-nosed lead bullet, no h.s. An experimental load with heavy bullet.

(b) Same case as above, 12.85mm bullet, either lighter than above or loaded deeper in the case, no h.s.

(c) Same case as above, 12.62mm bullet with no grooves showing, and a more pointed configuration than (a) or (b). No h.s.

(d) Blank for above in cut off 28mm case closed with cork wad. This blank may have chambered in the 12mm Lund carbine which used a 28mm case (See RFR 38). No h.s.

Comments

With their adoption of the Remington rolling block rifle in 1867, the Norwegians first used the ammunition jointly adopted with Sweden. This had a 42mm copper case (See RFR 79). However, the Norwegians chose to increase the power of the load by lengthening the case and powder capacity. The first two specimens shown above are evidently experimental loadings since the standard round used until about 1884 is that shown at (c). The 12.17mm identity of this cartridge refers to the bore diameter. Bullet diameters were somewhat larger in order to properly seal the bore and take the rifling.

This cartridge was also used in the experimental Krag-Petersson rifle, Model 1876.

RFR 79 – 12.17x42R Swedish
Remington, Model 1867

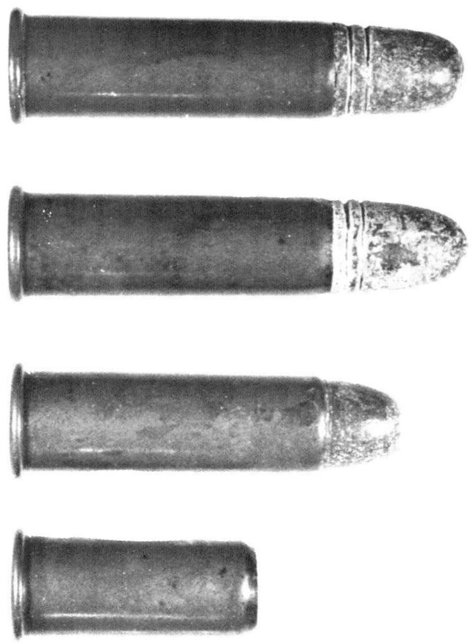

Description

(a) Copper case, 12.78mm two-groove round-nosed lead bullet, no h.s. An experimental loading.

(b) Same as above except bullet configuration more pointed. No h.s.

(c) Same case as above with no grooves showing on bullet. This is the standard load which Sweden adopted in 1867 along with Norway and maintained through use of the Remington rolling block rifle. No. h.s.

(d) Blank for above in 32mm case closed with cork wad. No h.s.

Comments

The Swedish Remington cartridge differs from the Norwegian only in its 2mm shorter case and a somewhat more blunt bullet. Centerfire cartridges of the same dimensions as the rimfires used by these two countries were loaded but only for use in sporting rifles.

RFR 80 – 11.7x42R (.460in) Danish Remington, Model 1867

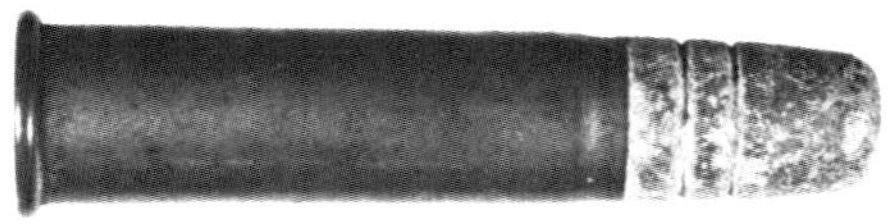

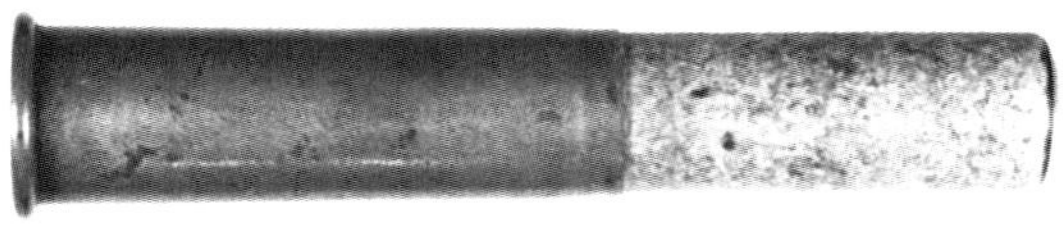

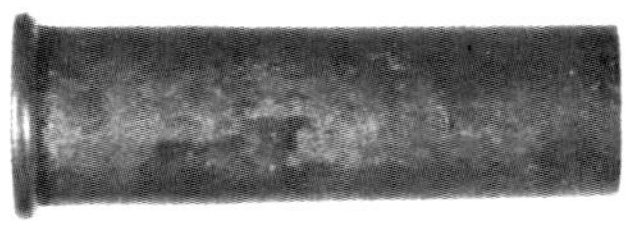

Description

(a) Copper case, 11.88mm two-groove flat-nosed lead bullet. Case shows a crimp at the mouth forming a slight shoulder. No h.s.

(b) Same case as above loaded with a paper-wrapped brass cylinder showing a lead filler at the crimped over nose. Projectile is loaded with incendiary composition for use as a flare. Impressed "T" headstamp.

(c) Same case as above closed with a cork wad. A blank charge with no h.s.

RFR 81 – 11.7x27R (.460in) Danish Remington Carbine, ca. 1870

Copper case with 12mm two-groove lead bullet. Case crimped into third groove. Impressed "T" headstamp.

Comments

The Danish Remington rifle cartridge was also loaded in the United States with round-nosed bullets. Winchester, in particular, is known to have manufactured ammunition for many foreign governments on contract so these may have been for military purposes.

The Danish Navy conducted trials with 100 experimental Krag-Petersson rifles, Model 1878 which presumably were chambered for the cartridge at RFR 80 above. The action was similar to the Peabody, with breechblock tipping down at the front to expose the chamber. The specimen illustrated below is shown courtesy the Danish National Museum–Tøjhusmuseet–Copenhagen.

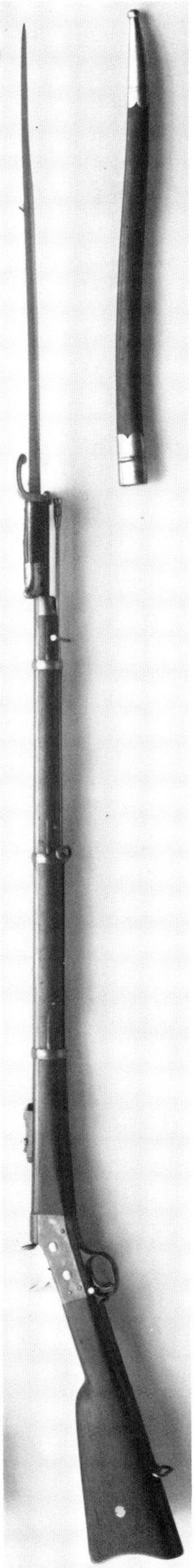

The Model 1867 Remington Rolling Block action used on the Danish, Norwegian, and Swedish rimfire rifles and carbines was essentially the same, the arms varying only in the details of construction required by the separate nations. The Remington was one of the simplest and strongest of all the single-shot actions. To operate, the hammer was drawn to cocked position and the breechblock rotated back by its lever. After the cartridge was inserted, the lever was thrust forward to rotate the block closed against it. When the hammer fell, its interior concealed portion rotated under the rear face of the breechblock and the action was locked when the hammer nose struck the firing pin. Shown here are the Danish infantry rifle, left, the action of the Danish carbine, above, and the same arm full length below. (Photos courtesy the Danish National Museum—Tøjhusmuseet—Copenhagen)

RFR 82 – 14.5x32.5R (.571in) Austrian
Waenzl, Model 1867

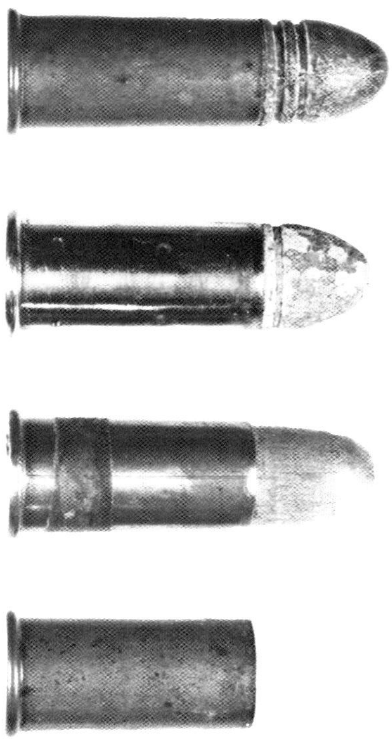

Description

(a) Copper case, 12mm two-groove flat-tipped lead bullet. Headstamp is an "N" with an arrow through it horizontally from right to left. A number of other headstamps exist (see below).

(b) *Brass* case loaded with a single-groove, flat-tipped bullet. No h.s. Since the Waenzl cartridge was loaded commercially, this may be one of those.

(c) Same case as (a) loaded with a wood bullet. No h.s. This is the military dummy round.

(d) Same case as (a) containing a blank charge closed with a card wad. No h.s. The blank also exists in brass case.

Comments

More headstamps have been found in the Waenzl cartridge than any other European rimfire. The intertwined "GR" and "GRW" were used by Georg Roth, Vienna (Wien), Austria, the 19th and early 20th century ammunition maker. The "K" was probably employed by Keller & Co., another Austrian cartridge manufacturer. Others may have been devised by merchants who sold the ammunition commercially and wanted their own identification thereon. (Drawings by M.E. Whiteley)

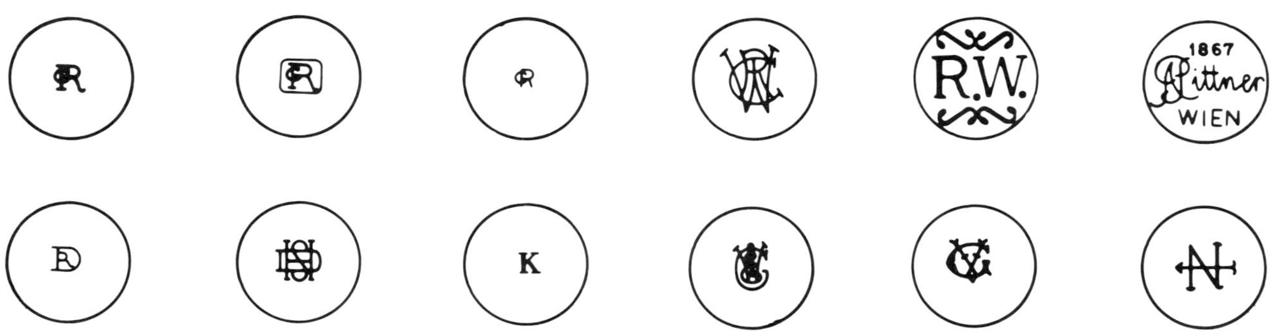

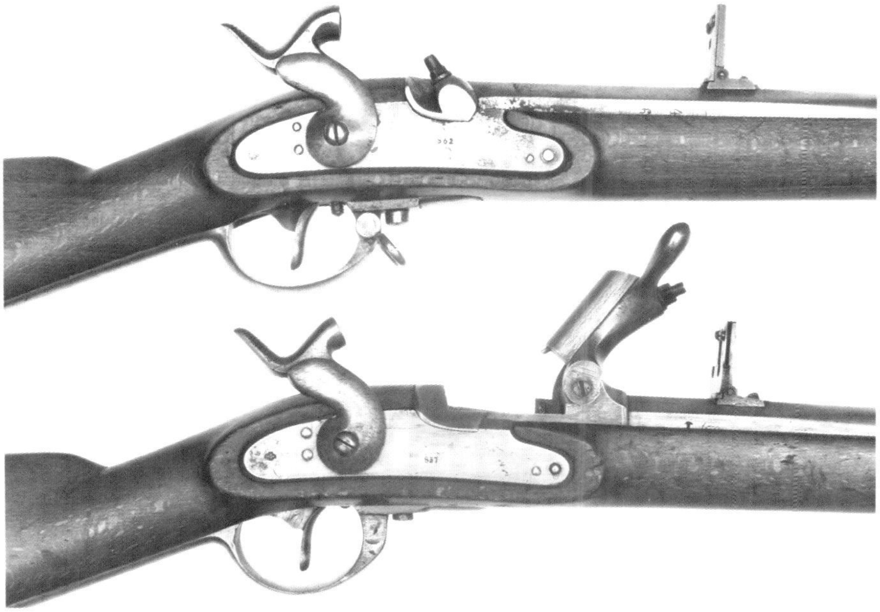

The Austrian Government adopted the breech-loading action of Vienna Gunmaker J. Waenzl (also found spelled Wanzel or Wanzl) to transform their muzzle-loading Lorenz rifles to rimfire. The breech end of the barrel was milled out on top, removing the percussion nipple and making room for a front-hinged breech block with rim firing pin fitted in a channel on the right side of the block with its head situated so that the original percussion hammer could be used. The lock plate was left undisturbed. An extracting device was included in the block to draw the cartridge case from the chamber when the breech block was thrown open and forward. Locking was accomplished by a bolt which entered a corresponding cavity in the rear of the breech block when the hammer descended. Shown are close up and full length views of the Austrian rifle before and after conversion to the Waenzl system. (Milwaukee Public Museum photos)

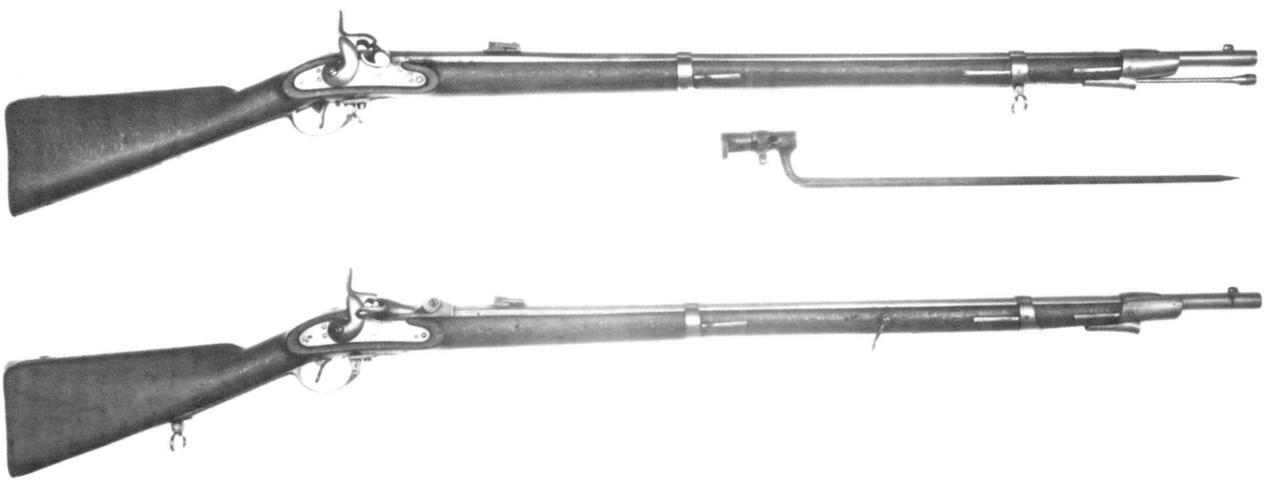

RFR 83 – 11.2x39.7R (.441in) Austrian Werndl or Remington Rifle, Model 1867

Description

Copper case, 11.63mm two-groove lead bullet. No h.s.

RFR 84 – 11.2x36R (.441in) Austrian Werndl or Remington Carbine, Model 1867

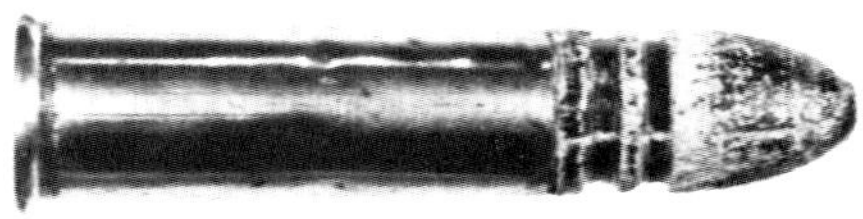

(a) Copper case, 11.63mm two-groove lead bullet. Headstamp is a large impressed "S."

(b) Same as above except shorter bullet with narrower grooves. Headstamp is a small "D" followed by a larger impressed "S."

(c) Sectioned specimen of above showing half of the "Pittner Wien" headstamp.

Comments

These cartridges were evidently developed for the trial Werndl system and Remington rolling block rifle and carbines submitted to the Austrian Government in 1867. The Werndl action was adopted. However, the arm and cartridges ultimately accepted were centerfire. The centerfire rifle load was of the same general configuration as RFR 83 above but the case was between 41 and 42mm long. The carbine centerfire case was nearly the same as the experimental rimfire.

Judging from the headstamps shown below found on specimens of the carbine cartridge, they must have been made in some quantity in Austria. The only one identified is the "Pittner Wien." Pittner is known to have made ammunition in Vienna; the same headstamp appears on the 14.5x32.5R Waenzl cartridge.

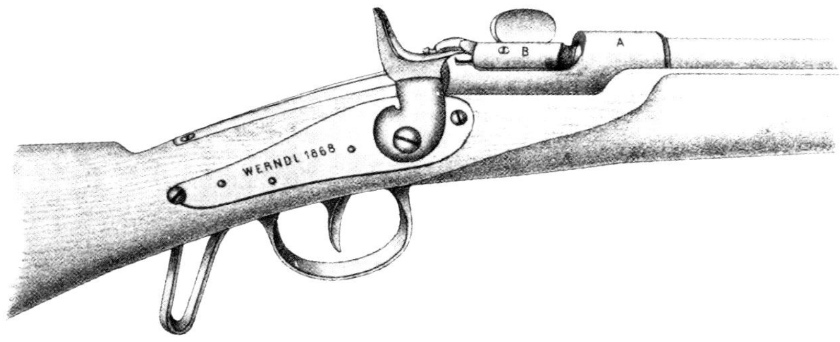

257

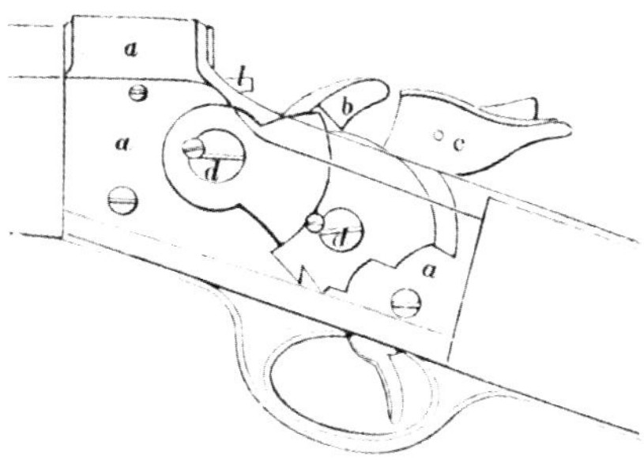

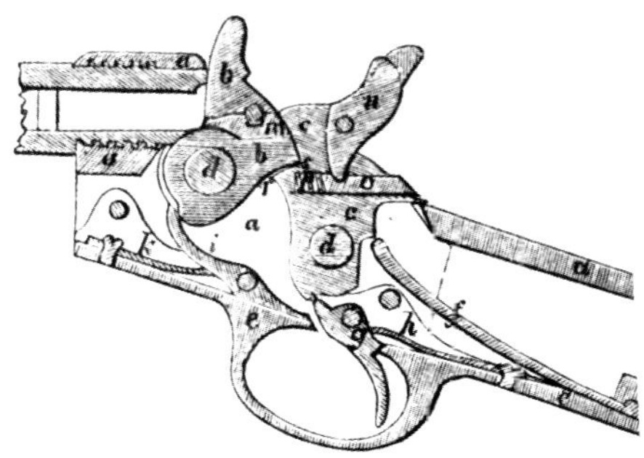

The Austrian Werndl action of 1868 rotated on a central pin on the same line as the rifle's bore. The underside was cut out so that turning the block by its operating handle clockwise brought the chamber into view, ready for loading. Turning it counterclockwise after inserting the cartridge readied the rifle for firing. The hammer struck a firing pin riding in an inclined passage bored through the breechblock.

The Remington Rolling Block action tested by the Austrian Government with the rimfire cartridges shown opposite was essentially the same in operating principle as that adopted by the Scandinavian countries with certain modifications, notably a safety hammer lever.

Neither arm was adopted in rimfire version. The Werndl was accepted using centerfire cartridges similar to the rimfire. (Drawings from Schmidt's "Les Armes A Feu Portatives," 1877 and "Kriegsfeuerwaffen" by Ellger, 1868)

Description

RFR 85 – 18mm (.709in) Swiss Milbank-Amsler

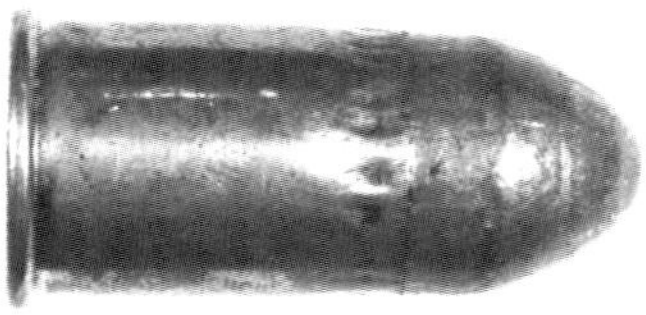

Brass 25.4mm case, 17.65mm bullet, headstamp is a raised "ST" in a depressed circle. Made by the Swiss firm Stahel, possibly a commercial load.

RFR 86 – 18mm (.709in) Swiss Milbank-Amsler, Extented Case Shot

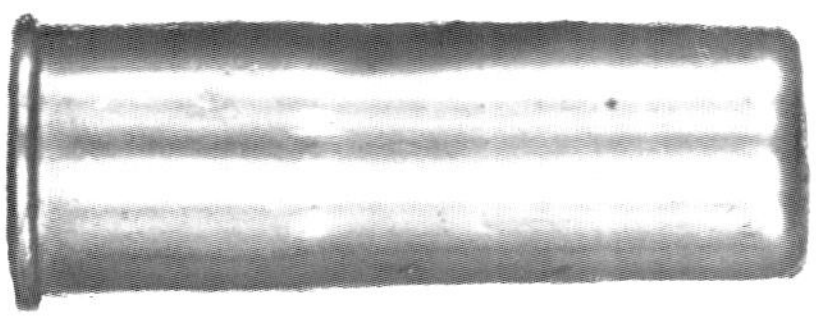

Brass 52.12mm case containing a shot load crimped over a card wad. Headstamp same as RFR 84.

RFR 87 – 18.3x24.6R (.721in) Unknown

Copper 24.6mm case, single groove 18.52mm lead bullet, no h.s. Case is distinctly necked midway. Presumed to be an experimental for a European transformed muzzle-loader.

RFR 88 – 10.4x39R (.409in) Swiss rimless unknown

Brass 39.1mm case, 10.79mm paper-patched lead bullet. Headstamp is raised "ST" in depressed circle (Swiss maker Stahel). Examination of interior base shows it to be a rimfire.

RFR 89 – 10.5x47R (.413in) Unknown

Copper 47.24mm case, 10.72mm single-groove lead bullet, no h.s. Case is slightly necked around the bullet. Presumed to be European since it differs in dimensions from any known American rimfire.

13 THE RIMFIRE MACHINEGUN CARTRIDGE

Dr. Richard Jordan Gatling is reported to have conceived the idea for a crank-operated machinegun in 1861. His gun was made with a cluster of barrels which rotated around a central axis. Ezra Ripley of Troy, New York patented a similar gun in October 1861. Gatling did not receive his until November 1862 but he never acknowledged any debt to Ripley.

The first of the Gatlings used steel cylinders with percussion nipple in one end and chamber in the other to accept a rifle-musket type paper cartridge. The cylinders fed by gravity to the firing barrel from the top of the gun.

The Gatling gun saw some service in the Union forces during the Civil War but was never given a trial by the military. In fact, it was often demonstrated in combat by Gatling's employees. Military thinking had not progressed with mechanical progress in weaponry so it remained a novelty.

It was not until January 1865 that Gatling submitted an improved model of his gun to the United States Army for trials. This model had four barrels and still used the paper cartridge loaded in a steel cylinder fitted with percussion nipple and fired by cap. However, in May of that year, Gatling patented another improvement that used a .58 caliber rimfire cartridge. These guns were first made by McWhinny and Rindge of Cincinnati and later by Cooper Manufacturing Company of Philadelphia, Pa. A 1-inch rimfire model evidently was tried out, probably by Frankford Arsenal. However, the only record of this is the few existing fired 1-inch rimfire cases in cartridge collections around the United States.

The .58 rimfire Gatling, and a centerfire cartridge of identical dimensions, were very short-lived. By 1866 these had been replaced by the .50 and 1-inch caliber centerfire models. Gatling apparently sold some of the rimfire models, but how many and where they went is not known. In any case, Union Metallic Cartridge Company loaded the ammunition into the 1870's.

References: "The Gatling Gun," Paul Wahl & Don Toppel, New York, 1965
"The Book of the Machine Gun," Major F.V. Longstaff and Captain A.H. Atterridge, 1917

RFMG 1 – 1in (25.4mm) U.S. Gatling Gun, 1865

Description

Copper 2.148in case, rim bearing an oblong firing pin mark. No h.s., no projectile.

RFMG 2 – 1in (25.4mm) U.S. Gatling Gun, (?) 1865

Copper 2.752in case, rim bearing a circular firing pin mark. No h.s., no projectile.

Head of upper specimen on left, lower specimen on the right.

Comments

The first rimfire Gatling cartridge shown above is one of three known to exist in collections. Frankford Arsenal experimented with the rimfire case for this hand cranked machinegun before abandoning rimfire ammunition and going to inside primed centerfire cartridges. It is believed that this is one of those experimentals.

The second specimen is unknown and presumed to be an experimental for the Gatling gun also. The case has the characteristics of ammunition made at that time by C.D. Leet, such as the .69in Roberts specimen previously shown. However, these statements await confirmation, or refutation.

RFMG 3 – .58in (14.73mm) U.S. Gatling, ca. 1865-1866

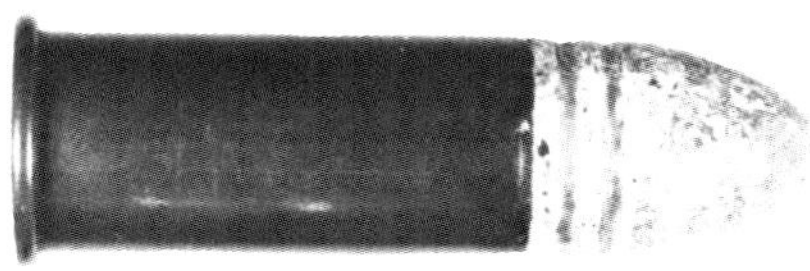

Description

Copper 1.37in case, .609in two-groove lead bullet, no h.s.

RFMG 4 – .58in (14.73mm) U.S. Gatling, ca. 1865-1866

Copper 1.12in case. .603in two-groove, more pointed, bullet. No h.s. Considered the intermediate case type.

RFMG 5 – .58in (14.73mm) U.S. Gatling, ca. 1865-1866

Copper .97in case, .596in two-groove bullet, no h.s. This is known as the Mountain Gatling cartridge.

Comments

Dr. Gatling's first rimfire model was made to use the same type of steel auxiliary chamber loaded with paper cartridge used in the first type, though it was bored all the way through and the head of the cartridge was exposed at the base end. However, he experienced difficulty getting these chambers lined up properly with the barrel as it came around to be fired. After unsuccessfully trying out an enlarged throat to the barrel to "funnel" the bullet into the bore, he ultimately developed a breech system that inserted the cartridge directly into the breech end of the barrel for firing and withdrew and extracted it in sequence.

This improved model 1865, using the rimfire cartridges shown here, and an experimental .58 centerfire identical in dimensions to the upper specimen, was soon superceded by centerfire models firing .50 and 1in ammunition.

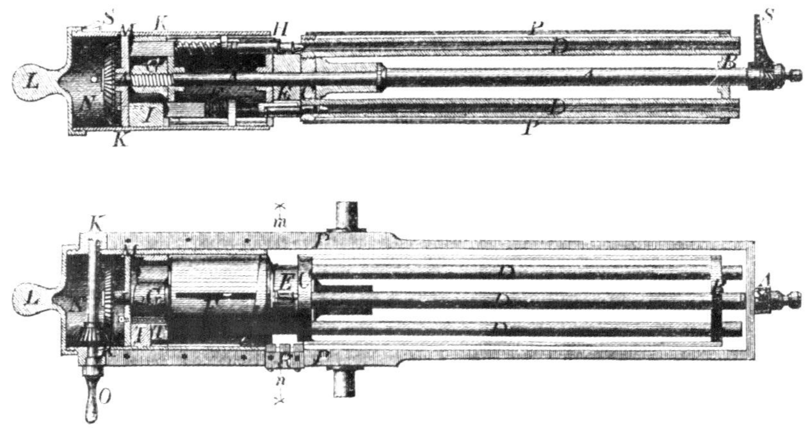

The original Gatling gun mechanism consisted of six rifled barrels mounted parallel to, and revolving around, a central axis. An open trough was milled in a carrier block behind each barrel. Six strikers were situated behind the breech, one behind each barrel and loading trough. This entire mechanism was rotated within the frame by a hand operated crank.

As the barrel, breech, and striker assembly rotated, a lug on the strikers was caused to bear on an inclined plane which first drew back the striker and then allowed it to fly forward under spring pressure to strike the percussion cap and fire that barrel. Continuing on in its sequence, that barrel rotated around until the fired chamber dropped out of its trough and to the ground below. At the same time, loaded cylinders fed by gravity action from a hopper on top of the breech mechanism and dropped one by one into the empty troughs as they came around. One complete revolution loaded, fired, and ejected six of the auxiliary chambers. After experimenting unsuccessfully with rimfire ammunition in bored-through auxiliary chambers, Dr. Gatling modified the action so that the metallic cartridge was chambered in the barrel itself and extracted after passing the firing stage. (Drawing from "Kriegsfeuerwaffen," Karl von Ellger, 1868)

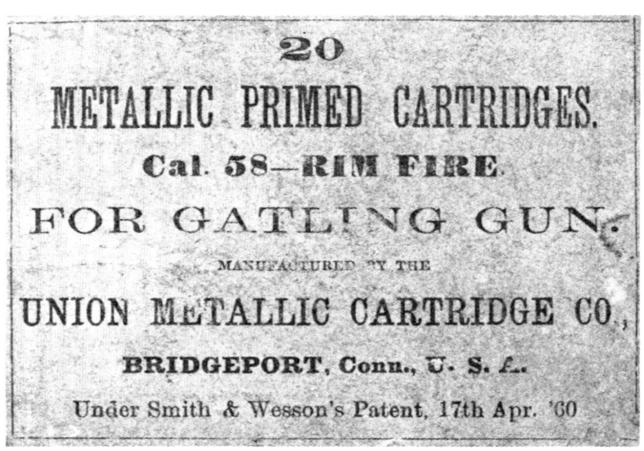

A FEW COMMENTS ON RIMFIRE CARTRIDGE DIMENSIONS

The nature of the foregoing presentation precluded showing a large number of specimens of a particular cartridge. In the Spencer group the collector will find much variation in case lengths, bullet diameters, and general appearance. This is so because so many companies made the ammunition from the Civil War into the 20th century. The condition of the copper case and the bullet influences dimension readings. Copper is often rough from deterioration and lead oxidizes, leaving a whitish surface which swells the diameter of the bullet.

Christopher Spencer identified his .56-56 cartridge by measuring the outside diameter of the case head and mouth. The rifle and carbine were .52in (bore diameter from land to opposite land). Groove diameters vary from .535 to .539. The bullets are nominally .54in though some measure larger. To further confuse matters, early makers often labeled the .56-56 Spencer boxes "No. 56." It is not unusual to see arms chambered for this cartridge incorrectly identified as .56 caliber. The Spencer sporting .56-46 rifles were .44in bore diameter (land to opposite land) and .455 groove diameter.
Some rimfires were made with heeled bullets; i.e., the base was reduced to fit the mouth of the cartridge case with the outside bullet diameter often larger than the case mouth.

Measuring the diameter of the bullet just outside the case mouth gives an approximate reflection of its caliber if that is the point of its maximum diameter. However, if the case is well crimped over the largest diameter of the bullet, this measurement does not serve to identify the maximum diameter of the bullet or indicate the caliber of the arms it was fired in.

Therefore, the bullet diameters reflected in the following charts attempt to give the actual maximum, even if this is within the case. Sometimes interpolation from the case measurements has been necessary to accomplish this.

The charts do not include all of the dummy and blank rounds shown because case length is already given in the description and such specimens are usually easily identified by referring to this and other case dimensions of loaded rounds, and their general appearance.

RIMFIRE ARMS AND CARTRIDGE DATA

First number in dimensions is inch, second millimeters. First under weight is grains, second grams.

No.	Caliber	Model & Date	Bullet Dia.	Bullet Weight	Powder Charge	Rim	Case Dimensions Head	Case Dimensions Mouth	Length	Total Length	Source
RFR 1	.42 10.67	U.S. Henry, 1861-	.443 11.25	210 14	25 1.62	.512 13	.444 11.28	.444 11.28	.832 21.12	1.363 34.60	Specimen—Lewis
RFR 2	.42 10.67	U.S. Henry, 1861-	.443 11.25	216 14	25 1.62	.520 13.21	.445 11.30	.445 11.30	.821 22.61	1.356 29.72	Specimen—Lewis
RFR 3(a)	.42 10.67	U.S. Henry, 1861-	.439 11.25	200 12.96	25 1.62	.514 13.06	.445 11.30	.445 11.30	.890 22.61	1.27 29.72	Specimen
RFR 3(b)	.42 10.67	U.S. Henry, 1861-	.439 11.25	200 12.96	25 1.62	.514 13.06	.445 11.30	.445 11.30	.895 22.73	1.362 34.59	Specimen—Lewis
RFR 3(c)	.42 10.67	U.S. Henry, ca. 1870 Austrian mfg.	.439 11.25			.518 13.16	.439 11.15	.439 11.15	.920 23.37	1.32 33.53	Specimen
RFR 4	.46 11.68	U.S. Remington, 1864	.453 11.51	296 19.18	41.5 2.69	.531 13.47	.458 11.63	.457 11.61	1.11 28.19	1.86 47.24	Specimen—Lewis
RFR 5	.44 11.18	U.S. Ballard and Wesson, 1859-1862	.438 11.13	205 13.28	25 1.62	.507 12.88	.441 11.20	.441 11.20	1.09 27.69	1.67 42.42	Specimen—Lewis
RFR 6	.36 9.14	Unknown, possibly Spencer, ca. 1860	.372 9.45			.509 12.93	.440 11.18	.387 9.83	.943 23.95	1.609 40.87	Specimen
RFR 7	.52 13.21	U.S. 56-56 Spencer, 1860	.542 13.77	450 29.20	40 2.59	.641 16.28	.559 14.20	.559 14.20	.89 22.61	1.65 41.91	Specimen—Lewis
RFR 8	.50 12.70	U.S. 56-50 Spencer, 1864	.520 13.21	350 22.68	45 2.92	.642 16.31	.561 14.25	.514 13.06	1.17 29.72	1.63 41.40	Specimen—Lewis
RFR 9	.50 12.70	U.S. 56-52 Spencer, ca. 1866	.530 13.46	386 25.01	45 2.91	.642 16.31	.559 14.20	.533 13.54	.950 24.13	1.66 42.16	Specimen—Williamson

RFR 10	.50 12.70	U.S. 56-52 Spencer, ca. 1866	.530 13.46	386 25.01	45 2.91	.642 16.31	.559 14.20	.532 13.51	.920 23.37	1.64 41.66	Specimen—Williamson
RFR 10(a)	.54 13.72	U.S. Ballard Old Model Rifle, 1861-	.548 13.92			.639 16.23	.560 14.22	.554 14.07	1.21 30.73	1.81 45.97	Specimen
RFR 10(b)	.54 13.72	U.S. Ballard Carbine, 1861	.543 13.89	427 27.67	53.6 3.47	.654 16.61	.578 14.68	.530 13.46	1.20 30.48	1.70 43.18	Specimen—Lewis
RFR 11(a)	.52 13.21	U.S. Sharps & Hankins Rifle, 1861-	.547 13.89	465 30.13	55 3.56	.635 16.13	.556 14.12	.556 14.12	1.18 29.97	1.80 45.72	Specimen—Lewis
RFR 11(b)	.52 13.21	U.S. Sharps & Hankins Carbine, 1861	.536 13.61	450 29.16	45 2.91	.640 16.26	.556 14.12	.542 13.77	1.15 29.21	1.67 42.42	Specimen—Lewis
RFR 11(c)	.52 13.21	U.S. Sharps & Hankins Carbine, 1861	.554 14.07			.643 16.33	.559 14.20	.559 14.20	1.02 25.91	1.65 41.91	Specimen—Lewis
RFR 12	.54 13.21	U.S. Joslyn Carbine, 1862-	.542 13.77	433 28.06	45 2.91	.641 16.28	.559 14.20	.559 14.20	.89 22.61	1.66 42.16	Specimen—Lewis
RFR 13	.58 14.73	U.S. Joslyn Carbine, 1862-	.600 15.24	380 24.62	45 2.91	.691 17.55	.612 15.54	.612 15.54	.985 25.02	1.23 31.24	Specimen—Lewis
RFR 14	.50 12.70	U.S. Ball & Lamson Carbine, 1864	.500 12.70			.640 16.26	.560 14.22	.476 12.09	.859 21.82	1.135 28.83	Specimen—Logan
RFR 15	.50 12.70	U.S. Warner Carbine, 1862	.515 13.08			.606 15.39	.529 13.44	.530 13.46	.86 21.84	1.52 38.61	Specimen
RFR 16	.54 13.72	U.S. Allen & Brand, 1864	.556 14.12	473 30.65	38.1 2.47	.640 16.26	.559 14.20	.559 14.20	.89 22.61	1.66 42.16	Specimen—Lewis
RFR 17	.50 12.70	U.S. Jenks Carbine, 1864	.539 13.69			.692 17.58	.615 15.62	.542 13.77	1.27 32.26	1.839 46.71	Specimen

RIMFIRE ARMS AND CARTRIDGE DATA

First number in dimensions is inch, second millimeters. First under weight is grains, second grams.

No.	Caliber	Model & Date	Bullet Dia.	Bullet Weight	Powder Charge	Rim	Case Dimensions Head	Case Dimensions Mouth	Length	Total Length	Source
RFR 18	.45 11.43	U.S. Peabody, 1862	.470 11.94	330 21.38	45 2.91	.644 16.36	.563 14.30	.484 12.29	1.357 34.47	1.761 44.73	Specimen
RFR 19	.45 11.43	U.S. Peabody, 1862	.470 11.94	420 27.21	60 3.89	.644 16.36	.561 14.25	.475 12.06	1.485 37.72	1.846 46.89	Specimen
RFR 20	.45 11.43	U.S. Peabody, 1862	.470 11.94	480 31.10	70 4.53	.646 16.41	.561 14.25	.475 12.06	1.702 43.23	2.060 52.34	Specimen
RFR 21	.45 11.43	U.S. Peabody, 1862	.470 11.94			.650 16.51	.566 14.38	.483 12.27	1.613 40.97	2.079 52.81	Specimen
RFR 22	.50 12.70	U.S. Peabody, 1862	.520 13.21	320 20.73	45 2.91	.644 16.36	.562 14.27	.508 12.90	1.162 29.51	1.601 40.66	Specimen
RFR 23	.50 12.70	U.S. Peabody, 1862	.520 13.21		50 3.24	.642 16.31	.561 14.25	.506 12.85	1.364 34.64	1.779 45.19	Specimen
RFR 24	.50 12.70	U.S. Peabody, 1862	.520 13.21	400 25.92	60 3.89	.642 16.31	.562 14.27	.506 12.85	1.477 37.51	1.922 48.82	Specimen—Lewis
RFR 25	.50 12.70	U.S. Peabody, 1862	.520 13.21		70 4.53	.642 16.31	.560 14.22	.506 12.85	1.665 42.04	2.084 52.93	Specimen
RFR 26	.45 11.43	U.S. Peabody, 1862	.470 11.94		70 4.53	.635 16.13	.563 14.30	.467 11.68	1.801 45.74	2.379 60.43	Specimen
RFR 27	.50 12.70	U.S. Peabody, 1862	.520 13.21		60 3.89	.643 16.33	.560 14.22	.515 13.08	1.463 37.16	2.206 56.03	Specimen

RFR 28	.50 12.70	U.S. Peabody, 1862	.520 13.21		60 3.89	.644 16.36	.563 14.30	.524 13.31	1.471 37.36	2.584 65.63	Specimen
RFR 29	.50 12.70	U.S. Joslyn, 1865 Conversion	.520 13.21		60 3.89	.642 16.31	.559 15.21	.527 13.38	1.46 37.08	1.93 49.02	Specimen
RFR 30(a)	.52 13.21	U.S. Sharps, 1865 Conversion	.529 13.44		70 4.53	.639 16.23	.562 14.27	.531 13.49	1.465 37.21	2.163 54.94	Specimen
RFR 30(b)	.52(?) 13.21	Unknown	.528 13.41			.646 16.41	.573 14.55	.530 13.46	1.541 39.13	2.25 57.15	Specimen
RFR 31	.54 13.72	U.S. Starr Carbine Conversion, 1867	.540 13.72			.645 16.38	.559 14.20	.556 14.12	1.03 26.16	1.69 42.93	Specimen
RFR 32	.58 14.73	U.S. Mont-Storm Conversion, 1858	.601 15.26			.732 18.59	.649 16.48	.618 15.70	1.202 30.53	1.90 48.26	Specimen
RFR 33	.69 17.53	U.S. Roberts Conversion, 1866	.712 18.08			.851 21.61	.742 18.85	.724 18.39	.740 18.80	1.47 37.34	Specimen
RFR 34	.69 17.53	U.S. Crittenden & Tibbals, 1866	.713 18.11	877 56.83	68 4.41	.835 21.21	.739 18.77	.739 18.77	1.07 27.18	1.75 41.91	Specimen
RFR 35	.58 14.73	U.S. Allin, Miller, Springfield, 1866	.590 14.99	500 32.40	60 3.89	.710 18.03	.630 16.00	.619 15.72	1.17 29.72	1.69 42.93	Specimen—Lewis
RFR 36	.58 14.73	U.S. "Long Musket," ca. 1866	.595 15.11			.700 17.78	.635 16.13	.605 15.37	1.389 35.28	1.864 47.34	Specimen
RFR 37	.58 14.73	U.S. Roberts, 1866	.602 15.29	480 31.10		.755 19.18	.651 16.53	.622 15.80	1.28 32.51	1.781 45.24	Specimen—Williamson

RIMFIRE ARMS AND CARTRIDGE DATA

First number in dimensions is inch, second millimeters. First under weight is grains, second grams.

No.	Caliber	Model & Date	Bullet Dia.	Bullet Weight	Powder Charge	Case Dimensions Rim	Case Dimensions Head	Case Dimensions Mouth	Length	Total Length	Source
RFR 38	.50 12.70	U.S. Springfield, 1866	.512 13.00	450 29.16	70 4.53	.655 16.64	.557 14.15	.532 13.51	1.72 43.69	2.19 55.62	Specimen
RFR 39	.44 11.18	U.S. Experimental, 1865-	.466 11.84	350 22.68	45 2.91	.585 14.86	.518 13.16	.460 11.68	1.136 28.85	2.03 51.56	Specimen—Lewis
RFR 40	.44 11.18	U.S. Experimental, 1865-	.465 11.81	300 19.44	40 2.59	.595 15.11	.518 13.16	.454 11.53	.998 25.35	1.687 42.85	Specimen—Lewis
RFR 41	.44 11.18	U.S. Experimental, 1865-	.471 11.96	350 22.68	40 2.59	.642 16.31	.561 13.11	.460 11.68	.881 22.38	1.722 43.74	Specimen—Lewis
RFR 42	.44 11.18	U.S. Experimental, 1865-	.472 11.99	300 19.44	40 2.59	.645 16.38	.558 14.17	.461 11.71	.881 22.38	1.579 40.11	Specimen—Lewis
RFR 43	.44 11.18	U.S. Experimental, 1865-	.459 11.66	500 32.40	45 2.91	.528 13.41	.444 11.28	.450 11.43	1.409 35.79	2.527 64.18	Specimen—Lewis
RFR 44	.44 11.18	U.S. Experimental, 1865-				.523 13.28	.455 11.56	.455 11.56	1.559 35.60		Specimen—Lewis
RFR 45	.44 11.18	U.S. Experimental, 1865-	.460 11.68	500 32.40	55 3.56	.523 13.28	.455 11.56	.452 11.48	1.675 42.54	2.804 71.22	Specimen—Lewis
RFR 46	.44 11.18	U.S. Experimental, 1865-	.469 11.91			.635 16.13	.558 14.17	.465 11.81	1.025 26.03	1.648 41.86	Specimen—Lewis
RFR 47	.44 11.18	U.S. Experimental, 1865-	.450 11.43			.641 16.28	.558 14.17	.454 11.53	1.025 26.03	1.757 44.63	Specimen—Lewis

RFR 48	.46 11.68	U.S. Experimental, 1865-	.469 11.91			.591 15.01	.512 13.00	.471 11.96	1.327 33.70	1.958 49.73	Specimen
RFR 49	.50 12.70	U.S. Experimental, 1865-	.518 13.16			.585 14.86	.513 13.03	.511 12.98	1.059 26.90	1.635 41.53	Specimen
RFR 50	.50 12.70	U.S. Experimental, 1865-	.512 13.00	500 32.40	50 3.24	.595 15.11	.520 13.21	.517 13.13	1.135 28.83	1.960 49.78	Specimen–Lewis
RFR 51	.50 12.70	U.S. Experimental, 1865-	.516 13.11	500 32.40	55 3.56	.594 15.09	.516 13.11	.518 13.16	1.247 31.67	2.153 54.69	Specimen–Lewis
RFR 52	.50 12.70	Unknown	.520(?) 13.21			.697 17.70	.608 15.44	.505 12.83	1.381 35.08	1.797 45.64	Specimen
RFR 53	.50 12.70	U.S. Experimental,	.515 13.08	480 31.10	65 4.21	.660 16.76	.582 14.78	.536 13.61	1.336 33.93	2.088 53.03	Specimen–Lewis
RFR 54	.41 10.4	Swiss, 1866	.425 10.79	312 20.22	60 3.89	.617 15.57	.540 13.72	.421 10.69	1.492 37.90	2.204 55.98	Specimen
RFR 55	.65 16.51	Unknown				.758 19.25	.661 16.79	.665 16.89	2.28 57.91		Specimen
RFR 56(a)	.50 12.70	U.S. Crispin, 1867-	.50 12.70			.665 16.89	.40 10.16	.497 12.62	1.50 38.10	1.94 49.28	Specimen
RFR 56(b)	.50 12.70	U.S. Crispin, 1867-	.50 12.70			.665 16.89	.40 10.16	.497 12.62	1.38 35.05	1.83 46.48	Specimen
RFR 56(c)	.50 12.70	U.S. Crispin, 1867-	.50 12.70			.665 16.89	.40 10.16	.495 12.57	1.34 34.04	1.70 43.18	Specimen
RFR 57	.52 13.21	Unknown	.538 13.67			.663 16.84	.571 14.50	.543 13.79	1.786 45.36	2.310 58.67	Specimen

RIMFIRE ARMS AND CARTRIDGE DATA

First number in dimensions is inch, second millimeters. First under weights is grains, second grams.

No.	Caliber	Model & Date	Bullet Dia.	Bullet Weight	Powder Charge	Case Dimensions Rim	Case Dimensions Head	Case Dimensions Mouth	Length	Total Length	Source
RFR 58	.50 12.70	U.S. Unknown, Presumed Burton, ca. 1866(?)	.520 13.21		48(?) 3.11	.565 14.45	.570 14.48	.514 13.05	1.131 28.73	1.581 40.16	Specimen
RFR 59	.44 11.18	U.S. Dupee Patent by Winchester, ca. 1870	.454 11.53			.509 12.93	.509 12.93	.456 11.58	1.089 48.00	1.514 38.45	Specimen
RFR 60	.52 13.21	U.S. Spencer, 1870-	.550 13.97	350 22.68		.640 16.26	.557 14.15	.557 14.15	.872 22.15	1.512 38.40	Specimen—Williamson
RFR 61	.50 12.70	U.S. Spencer, 1870-	.493 12.52	350 22.68		.642 16.31	.556 14.12	.527 13.38	1.04 26.42	1.62 41.15	Specimen
RFR 62	.50 12.70	U.S. Spencer, 1870-				.639 16.23	.559 14.21	.520 13.21	1.54 39.12	1.54 39.12	Specimen
RFR 63	.455 11.56 (grooves)	U.S. Spencer, 1870-	.474 12.04			.646 16.41	.559 14.21	.491 12.47	.910 23.11	1.610 41.15	Specimen
RFR 64	.455 11.56 (grooves)	U.S. Spencer, 1870-	.469 11.91	330 26.54		.639 16.23	.558 14.17	.480 12.19	1.04 26.42	1.57 39.88	Specimen
RFR 65	.46 11.68	U.S. Extra Long, ca. 1870's-	.461 11.71	305 19.76	57 3.69	.520 13.21	.457 11.61	.456 11.58	1.533 38.94	2.26 57.40	Specimen—Suydam
RFR 66	.46 11.68	U.S. Long, ca. 1865-	.453 11.51	300 19.44	40 2.59	.523 13.28	.456 11.58	.455 11.56	1.251 31.77	1.88 47.75	Specimen—Suydam
RFR 67	.44 11.18	U.S. Extra Long, ca. 1870's-	.442 11.23	218 14.13	46 2.98	.510 12.95	.440 11.18	.440 11.18	1.486 37.74	2.06 52.32	Specimen—Williamson
RFR 68	.44 11.18	U.S. Ballard, 1861-	.440 11.18	220 14.25	28 1.81	.514 13.05	.441 11.20	.441 11.20	1.12 28.45	1.652 41.96	Specimen—Williamson

RFR 69	.44 11.18	U.S. Long, 1865-	.440 11.18	220 14.25	28 1.81	.508 12.90	.438 11.12	.438 11.12	.931 23.65	1.473 37.41	Specimen—Williamson
RFR 70	.38 9.65	U.S. Extra Long, ca. 1870's-	.380 9.65	148 9.59	38 2.46	.438 11.07	.378 9.60	.378 9.60	1.47 37.34	1.84 46.74	Specimen—Suydam
RFR 71	.32 8.13	U.S. Extra Long, ca. 1870's	.320 8.13	90 5.83	20 1.29	.367 9.32	.318 8.08	.318 8.08	1.10 27.94	1.52 38.61	Specimen—Suydam
RFR 72(a)	.709 18.00	Swiss Milbank-Amsler, 1867	.705 17.91	617 40	69 4.50	.815 20.70	.730 18.54	.705 17.91	1.00 25.40	1.60 40.64	Specimen
RFR 72(b)	.709 18.00	Swiss Milbank-Amsler, 1867	.706 17.94	617 40	69 4.50	.847 21.51	.740 18.80	.729 18.52	1.00 25.40	1.46 37.08	Specimen
RFR 73(a)	.409 10.4	Swiss Milbank-Amsler, 1867	.401 10.18	287 18.6	56 3.65	.620 15.75	.539 13.69	.440 11.18	1.50 38.10	2.30 58.42	Specimen
RFR 73(b)	.409 10.4	Swiss Milbank-Amsler, 1867	.420 10.67	295 19.1	56 3.65	.623 15.82	.540 13.72	.418 10.16	1.50 38.10	2.22 56.39	Specimen
RFR 74(a)	.409 10.4	Swiss Vetterli, 1871-	.415 10.54	312 20.2	56 3.65	.619 15.72	.539 13.69	.446 11.33	1.50 38.10	2.22 56.39	Specimen
RFR 74(b)	.409 10.4	Swiss Vetterli, 1871-	.415 10.54	315 20.4	56 3.65	.622 15.80	.538 13.66	.442 11.23	1.50 38.10	2.22 56.39	Specimen
RFR 74(c)	.409 10.4	Swiss Vetterli, 1871-	.438 11.12		56 3.65	.623 15.82	.538 13.66	.426 10.82	1.50 38.10	2.23 56.42	Specimen
RFR 74(d)	.409 10.4	Swiss Vetterli, 1871-	.423 10.74			.623 15.82	.538 13.66	.428 10.88	1.48 38.04	1.70 43.18	Specimen
RFR 74(e)	.409 10.4	Swiss Vetterli, 1871-	.415 10.54			.623 15.82	.539 13.69	.428 10.88	1.50 38.10	2.19 56.30	Specimen
RFR 74(f)	.409 10.4	Swiss Vetterli, 1871-	.422 10.71			.623 15.72	.540 13.72	.436 11.07	1.50 38.10	2.21 56.36	Specimen

RIMFIRE ARMS AND CARTRIDGE DATA

First number in dimensions is inch, second millimeters. First under weight is grains, second grams.

No.	Caliber	Model & Date	Bullet Dia.	Bullet Weight	Powder Charge	Rim	Case Dimensions Head	Case Dimensions Mouth	Length	Total Length	Source
RFR 75	.669 17.00	Norwegian Lund, 1867	.690 17.56			.823 20.90	.723 18.36	.712 18.08	1.24 31.50	1.74 44.20	Specimen
RFR 76(a)	.472 11.99	Norwegian Lund, Carbine, 1869	.498 12.65			.632 16.05	.553 14.04	.537 13.64	1.16 29.46	1.56 39.62	Specimen
RFR 76(b)	.472 11.99	Norwegian Lund Carbine, 1869	.452 11.48			.632 16.05	.553 14.04	.532 13.51	1.13 28.70	1.59 40.39	Specimen
RFR 77(a)	.689 17.50	Danish Snider, 1865	.685 17.40			.792 20.12	.706 17.93	.703 17.86	1.09 27.69	1.57 39.88	Specimen
RFR 77(b)	.689 17.50	Danish Snider, 1865	.632 16.05			.775 19.68	.713 18.11	.694 17.63	1.07 27.18	1.72 43.69	Specimen
RFR 78(a)	.479 12.17	Norwegian Remington, 1871	.508 12.90			.632 16.05	.551 13.99	.532 13.51	1.72 43.69	2.50 63.50	Specimen
RFR 78(b)	.479 12.17	Norwegian Remington, 1871	.506 12.85			.628 15.95	.552 14.02	.532 13.51	1.72 43.69	2.39 60.71	Specimen
RFR 78(c)	.479 12.17	Norwegian Remington, 1871	.506 12.85			.629 15.98	.552 14.02	.532 13.51	1.72 43.69	2.25 57.15	Specimen
RFR 79(a)	.479 12.17	Swedish Remington, 1867	.506 12.85			.637 16.18	.552 14.02	.535 13.59	1.65 41.91	2.33 59.18	Specimen
RFR 79(b)	.479 12.17	Swedish Remington, 1867	.506 12.85			.629 15.98	.551 13.99	.534 13.56	1.65 41.91	2.35 59.69	Specimen
RFR 79(c)	.479 12.17	Swedish Remington, 1867	.506 12.85			.633 16.08	.544 13.82	.536 13.62	1.60 40.64	2.05 52.07	Specimen

RFR 80(a)	.460 11.70	Danish Remington Rifle, 1867	.465 11.81	.585 14.86	.502 12.75	.481 12.22	1.64 41.66	2.32 58.93	Specimen
RFR 80(b)	.460 11.70	Danish Remington Rifle, 1867	.448 11.38	.584 14.83	.503 12.78	.472 11.99	1.62 41.15	2.81 71.37	Specimen
RFR 81	.460 11.70	Danish Remington Carbine, ca. 1870	.465 11.81	.586 14.89	.502 12.75	.486 12.34	1.08 27.43	1.60 40.64	Specimen
RFR 82(a)	.571 14.50	Austrian Waenzl, 1867	.576 14.63	.689 17.50	.607 15.42	.589 14.96	1.28 32.51	1.97 50.04	Specimen
RFR 82(b)	.571 14.50	Austrian Waenzl, 1867	.571 14.50	.699 17.75	.607 15.42	.591 15.01	1.30 33.02	1.87 47.50	Specimen
RFR 82(c)	.571 14.50	Austrian Waenzl, 1867	.532 14.50	.703 17.86	.609 15.47	.589 14.96	1.25 31.75	1.90 48.26	Specimen
RFR 83	.441 11.20	Austrian Werndl & Remington Rifle, 1867	.463 11.76	.572 14.53	.490 12.45	.473 12.01	1.56 39.70	2.29 58.20	Specimen
RFR 84(a)	.441 11.20	Austrian Werndl & Remington Carbine, 1867	.463 11.76	.572 14.53	.490 12.45	.473 12.01	1.40 36.07	2.23 56.64	Specimen
RFR 84(b)	.441 11.20	Austrian Werndl & Remington Carbine, 1867	.463 11.76	.572 14.53	.490 12.45	.473 12.01	1.40 36.07	2.18 55.37	Specimen
RFR 85	.709 18.00	Swiss Milbank-Amsler	.695 17.65	.814 20.67	.727 18.46	.698 17.73	1.00 25.40	1.63 41.40	Specimen
RFR 86	.709 18.00	Swiss Milbank-Amsler		.815 20.70	.726 18.44	.628 15.95	2.052 52.12	2.052 52.12	Specimen

RIMFIRE ARMS AND CARTRIDGE DATA

First number in dimensions is inch, second millimeters. First under weight is grains, second grams.

No.	Caliber	Model & Date	Bullet Dia.	Bullet Weight	Powder Charge	Case Dimensions Rim	Case Dimensions Head	Case Dimensions Mouth	Length	Total Length	Source
RFR 87	.721(?) 18.31	Unknown	.729 18.52			.833 21.16	.767 19.48	.721 18.31	.970 24.64	1.51 38.35	Specimen
RFR 88	.409 10.4	Swiss Rimless Experimental	.425 10.79			.534 13.56	.540 13.72	.439 11.15	1.536 39.01	2.237 56.82	Specimen
RFR 89	.413(?) 10.50	Unknown	.422 10.72			.573 14.55	.489 12.42	.443 11.25	1.86 47.24	2.50 63.50	Specimen
RFMG 1	1-in. 25.40	U.S. Gatling, 1865	.98 24.89	3,500 227		1.16 29.46	1.43 36.32	1.29 32.77	2.15 54.61		Specimen—Lewis (Ord. Memo 14)
RFMG 2	1-in. 25.40	U.S. Gatling, 1865				1.144 28.96	1.040 26.42	1.035 26.29	2.752 69.90		Specimen
RFMG 3	.58 14.73	U.S. Gatling, 1865	.609 15.47			.704 17.88	.613 15.57	.613 15.57	1.37 34.80	2.15 54.61	Specimen
RFMG 4	.58 14.73	U.S. Gatling, 1865	.603 15.32			.682 17.32	.607 15.42	.609 15.47	1.12 28.45	1.97 50.04	Specimen
RFMG 5	.58 14.73	U.S. Gatling, 1865	.596 15.14			.709 18.00	.614 15.59	.611 15.52	.970 24.64	1.80 45.72	Specimen

APPENDIX A

The U.S., Canadian, and French box labels on the following pages trace the manufacture of Spencer military and sporting ammunition from the early 1860's well into the 20th century. Information, type size and style, as well as label design varied with the same maker over the years so no two presented here are identical.

Crittenden & Tibbals Manufacturing Company of South Coventry, Connecticut was formed about 1850 to make percussion caps. The Hunt "Rocket Ball" was also made by this firm. C & T, and C.D. Leet of Springfield, Massachusetts loaded much of the small arms ammunition used by the Union during the Civil War.

After the War, the two companies were bought by Schuyler, Hartley and Graham, arms dealers of New York. Marcellus Hartley served as a European arms buyer for the North during the war and masterminded the formation of Union Metallic Cartridge & Cap Company, using machinery of C & T and C.D. Leet which he moved to Bridgeport, Conn. The new firm was incorporated as Union Metallic Cartridge Company in August 1867. The labels shown here date up to at least 1911 when Remington Arms Company and U.M.C. were joined. Thereafter, box labels were Remington-U.M.C.

Sharps & Hankins as well as Bacon Manufacturing Company ceased ammunition production with the close of the Civil War. The United States Cartridge Company label dates well after the firm's 1869 incorporation date, and probably between 1880 and 1900. Winchester bought out this maker in 1926.

Winchester Repeating Arms Company was formed in 1866 and soon thereafter expanded to ammunition production. The box labels here date from 1873 into the early 20th century.

Dominion Cartridge Company of Canada formed at Brownsburg, Quebec, in 1886 made Spencer ammunition from around that date well into the 20th century, possibly up to 1928 when it became Dominion Ammunition Division of Canadian Industries Limited (C.I.L.). American Metallic Cartridge Company, successor to Phoenix Metallic Cartridge Company formed in 1872, was bought by U.M.C. in 1890 so the box label shown probably was printed during the 1880's. (Box label photographs courtesy George Kass)

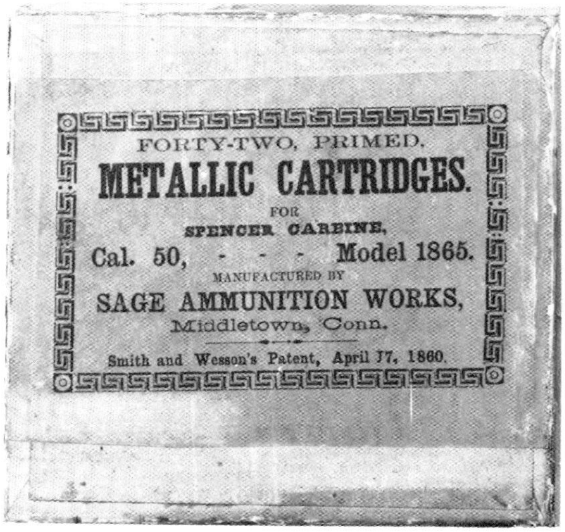
FORTY-TWO, PRIMED.
METALLIC CARTRIDGES.
FOR
SPENCER CARBINE,
Cal. 50, - - - Model 1865.
MANUFACTURED BY
SAGE AMMUNITION WORKS,
Middletown, Conn.
Smith and Wesson's Patent, April 17, 1860.

FORTY-TWO, PRIMED
METALLIC CARTRIDGES.
FOR
SPENCER CARBINE
Cal. 50, - - - - - Model 1865.
MANUFACTURED BY
SAGE AMMUNITION WORKS,
Middletown, Conn.

FORTY-TWO, PRIMED
METALLIC CARTRIDGES
FOR
SPENCER CARBINE,
Cal. 50, - - Model 1865.
MANUFACTURED BY
SAGE AMMUNITION WORKS,
Middletown, Conn.

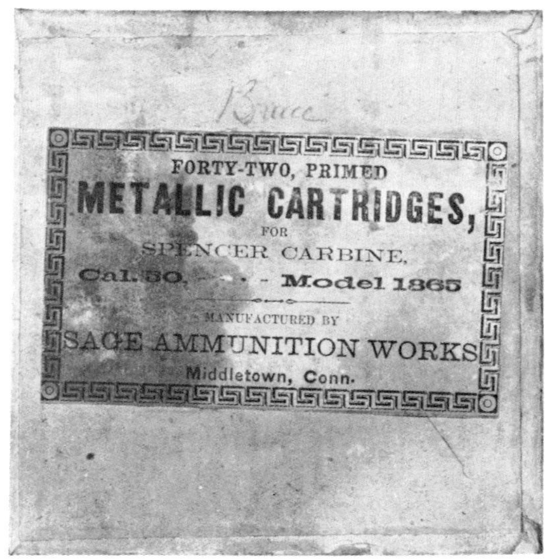
Bruce
FORTY-TWO, PRIMED
METALLIC CARTRIDGES,
FOR
SPENCER CARBINE,
Cal. 50, - - - Model 1865
MANUFACTURED BY
SAGE AMMUNITION WORKS
Middletown, Conn.

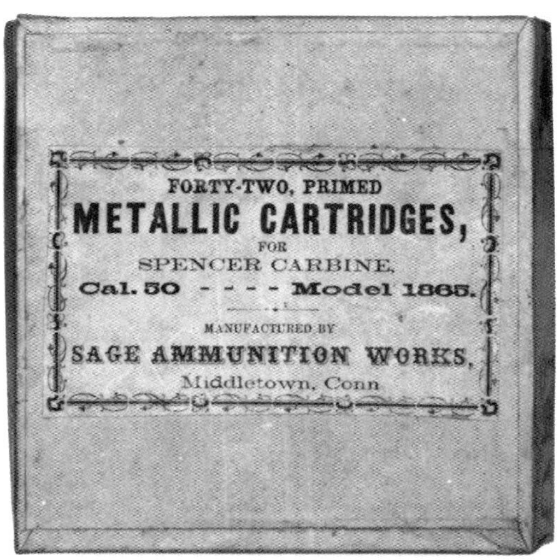
FORTY-TWO, PRIMED
METALLIC CARTRIDGES,
FOR
SPENCER CARBINE,
Cal. 50 - - - - Model 1865.
MANUFACTURED BY
SAGE AMMUNITION WORKS,
Middletown, Conn

FORTY-TWO, PRIMED,
METALLIC CARTRIDGES.
FOR
SPENCER CARBINE,
Cal. 50, - - - Model 1865.
MANUFACTURED BY
SAGE AMMUNITION WORKS,
Middletown, Conn.

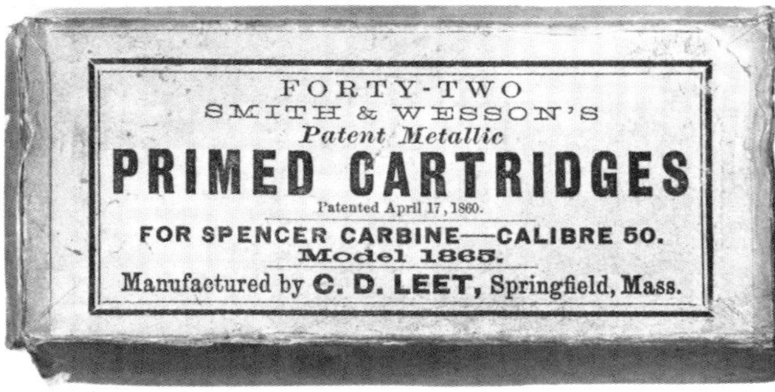
FORTY-TWO
SMITH & WESSON'S
Patent Metallic
PRIMED CARTRIDGES
Patented April 17, 1860.
FOR SPENCER CARBINE—CALIBRE 50.
Model 1865.
Manufactured by C. D. LEET, Springfield, Mass.

FORTY-TWO
SMITH & WESSON'S
PATENT METALLIC
PRIMED CARTRIDGES
Patented April 17 1860.
For Spencer Carbine—Calibre 50.
MODEL 1865.
Manufactured by C. D. LEET, Springfield, Mass.

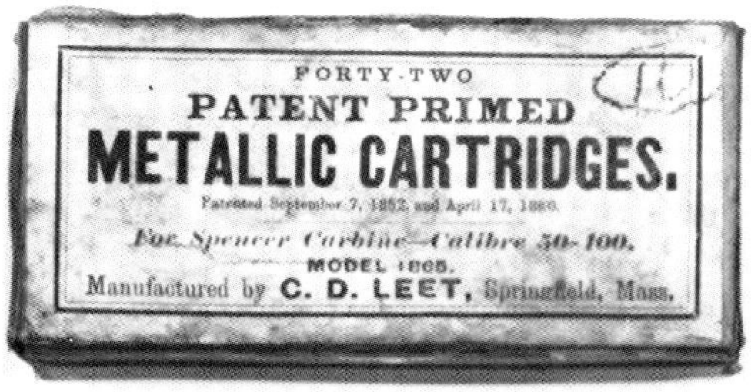
FORTY-TWO
PATENT PRIMED
METALLIC CARTRIDGES.
For Spencer Carbine—Calibre 50-100.
MODEL 1865.
Manufactured by C. D. LEET, Springfield, Mass.

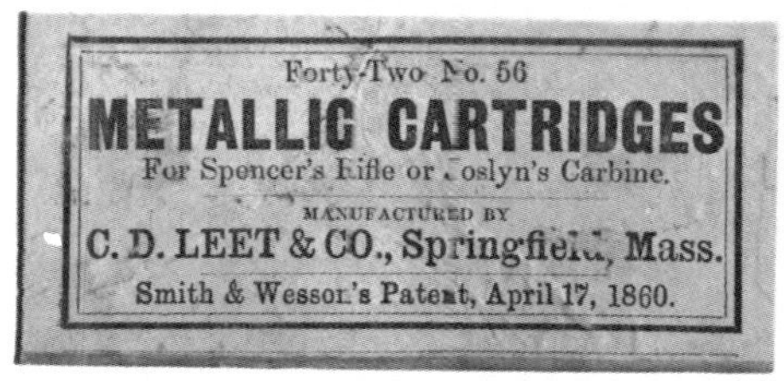
METALLIC CARTRIDGES
MANUFACTURED BY
Smith & Wesson's Patent, April 17, 1860.

CARTOUCHES
pour
MOUSQUETONS
TYPE N° 1

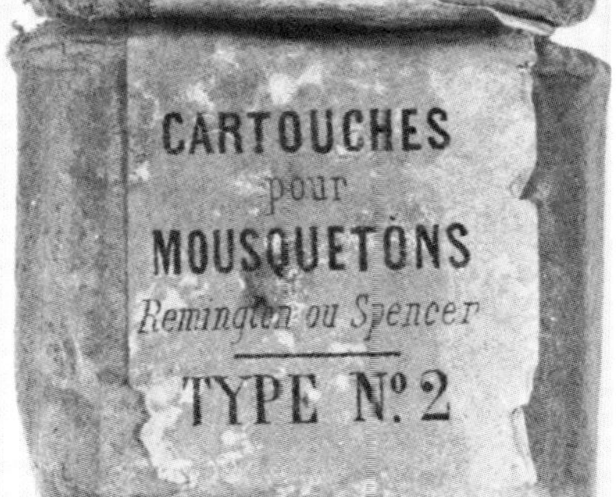
CARTOUCHES
pour
MOUSQUETONS
Remington ou Spencer
TYPE N° 2

25 No. 56
Metallic Cartridges
FOR JOSLYN'S CARBINES.
Manufactured by
C. D. LEET & CO.,
SPRINGFIELD, MASS.

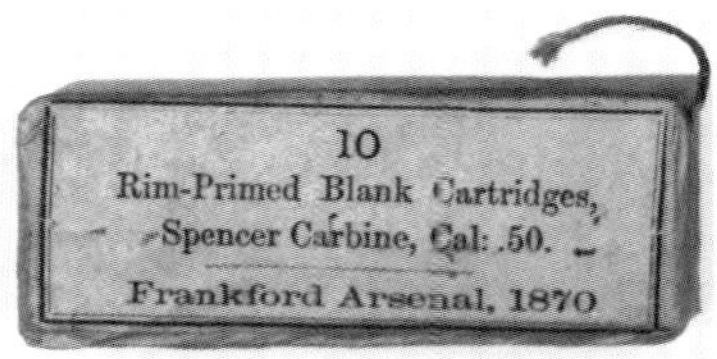
10
Rim-Primed Blank Cartridges,
Spencer Carbine, Cal. .50.
Frankford Arsenal, 1870

50, No. 38,
METALLIC WATER-PROOF
CARTRIDGES,
NAVY SIZE.
SOUTH COVENTRY, Conn.
Also, ORDNANCE AND SMALL ARMS PRIMINGS.

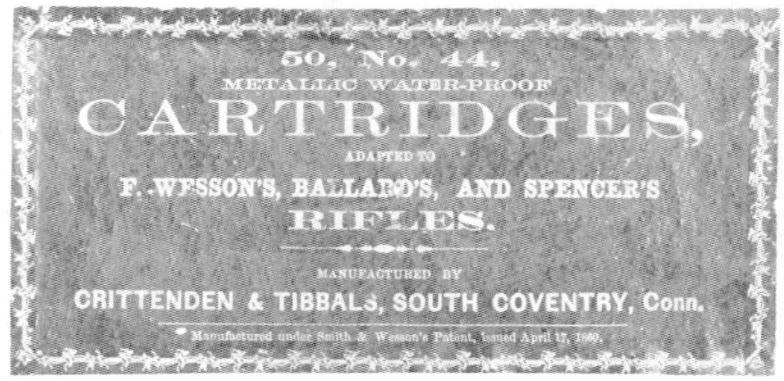
50, No. 44,
METALLIC WATER-PROOF
CARTRIDGES,
ADAPTED TO
WESSON'S, AND SPENCER'S
RIFLES.
MANUFACTURED BY
CRITTENDEN & TIBBALS, SOUTH COVENTRY, Conn.

50
CALIBRE: 56/56 RIM-FIRE
Metallic Cartridges,
SPENCER MAGAZINE RIFLE,
U. S. CARTRIDGE COMPANY, LOWELL, MASS.

42
METALLIC CARTRIDGES,
FOR THE
SPENCER & JOSLYN CARBINE,
No. 56 Navy and Infantry Size.
CRITTENDEN & TIBBALS,
South Coventry, Conn., U. S.

.56-56 SPENCER
25 RIM FIRE CARTRIDGES
REMINGTON
UMC
REMINGTON ARMS-UNION METALLIC CARTRIDGE Co
UNION METALLIC CARTRIDGE WORKS

.56-50 Spencer
25 RIM FIRE CARTRIDGES
SHOT
REMINGTON ARMS-UNION METALLIC CARTRIDGE Co
UNION METALLIC CARTRIDGE WORKS

CARTRIDGES,
NAVY SIZE.
BACON MANUFACTURING CO., NORWICH, CONN.

.56-52 Spence
25 RIM FIRE CARTRIDG
SHOT CARTRIDGES
SPENCER
UMC
REMINGTON ARMS-UNION METALLIC CARTRIDGE Co
UNION METALLIC CARTRIDGE WORKS

.56-46 Spencer
25 RIM FIRE CARTRIDGES
REMINGTON
UMC
REMINGTON ARMS-UNION METALLIC CARTRIDGE Co

METALLIC CARTRIDGES
For Spencer's Rifle, or Joslyn's Carbine.
PERCUSSIONED BY SHARPS' PROCESS, PATENTED DEC. 1, 1863.
MANUFACTURED BY
PHILADELPHIA.

25
.56-52
PRIMED SHELLS
WINCHESTER
REPEATING ARMS CO.,
NEW HAVEN, CONN., U. S. A.
FOR
SPENCER
ARMS.
STETSON'S PAT.
OCT. 24, '74.
25
"WINCHESTER"
TRADE MARK.
25
.56-50 CALIBER CARTRIDGES
SHOT CARTRIDGES.
FOR SP ARMS.
NE AVEN, CONN., U. S. A.
25
.56-56
SHOT CARTRIDGES.
WINCHESTER
REPEATING ARMS CO.
NEW HAVEN, CONN., U. S. A.
SPENCER
ARMS.
STETSON'S PAT.
OCT. 24, '74.
25
.56-52
METALLIC CARTRIDGES.
Manufactured by the
WINCHESTER
REPEATING ARMS CO.,
NEW HAVEN, CONN., U. S. A.
FOR
SPENCER
ARMS.
TRADE
MARK.
STETSON'S PAT.
OCT. 24. '74.
25
.56-50
METALLIC CARTRIDGES.
Manufactured by the
WINCHESTER
REPEATING ARMS CO.
NEW HAVEN, CONN., U. S. A.
FOR
SPENCER
ARMS.
TRADE
MARK.
STETSON'S PAT.
OCT. 24, '74.
25
.56-46
METALLIC CARTRIDGES.
WINCHESTER
REPEATING ARMS CO.
SPENCER
STETSON'S PAT.
25
.56-56
METALLIC CARTRIDGES.
Manufactured by the
WINCHESTER
REPEATING ARMS CO.
NEW HAVEN, CONN., U. S. A.
FOR
SPENCER
ARMS.
Trade Mark
Stetson's Pat.
OCT. 31, 1871.
Registered in U. S. A.
25
.56-56
SHOT CARTRIDGES.
METALLIC CARTRIDGES.
Manufactured by the
WINCHESTER
REPEATING ARMS CO.
NEW HAVEN, CONN., U. S. A.
FOR
SPENCER
ARMS.
Trade Mark
Stetson's Pat.
OCT. 31, 1871.
Registered in U. S. A.
25
.56-50
BLANK CARTRIDGES.
WINCHESTER
REPEATING ARMS CO.
NEW HAVEN, CONN., U. S. A.
FOR
SPENCER
ARMS.
TRADE
MARK.
STETSON'S PAT.
OCT. 24, '74.

25
Metallic Cartridges
Dominion Cartridge Co.
Montreal, Canada
25
Metallic Cartridges
25
SHOT CARTRIDGES,
CALIBRE 56-52.
MANUFACTURED BY THE
American Metallic Cartridge Co.
South Coventry, Conn.
25
56-46
Metallic Cartridges
Dominion Cartridge Co., Ltd.
MONTREAL
56-56 R.F. BLACK
ADAPTED TO
BALLARD,
JOSLYN
AND
SPENCER
CARBINES
MANUFACTURED BY THE
DOMINION CARTRIDGE CO.
MONTREAL, CANADA. LIMITED
25
WINCHESTER
REPEATING ARMS CO.
FOR
SPENCER
ARMS.
STETSON'S PAT.
OCT. 24, '74.
25
Metallic Cartridges
Dominion Cartridge Co., Ltd.
MONTREAL
25 "WINCHESTER" 25
.56-50 CALIBER CARTRIDGES
RIM FIRE
FOR SPENCER ARMS.
Stetson's Patent Oct. 24, 1874.
WINCHESTER REPEATING ARMS CO.
NEW HAVEN, CONN., U. S. A.
25 56-56
METALLIC CARTRIDGES.
WINCHESTER
REPEATING ARMS CO.
NEW HAVEN, CONN., U. S. A.
FOR
SPENCER
ARMS.
STETSON'S PAT.
OCT. 24, '74.
56-52
100
42 CARTRIDGES
WINCHESTER
REPEATING ARMS CO.
NEW HAVEN CONN. U.S.A.
SPENCER ARMS.
56-46
100
42 CARTRIDGES
SPENCER SPORTING
MANUFACTURED BY THE
WINCHESTER REPEATING ARMS CO.
NEW HAVEN, CONN.
U. S. A.

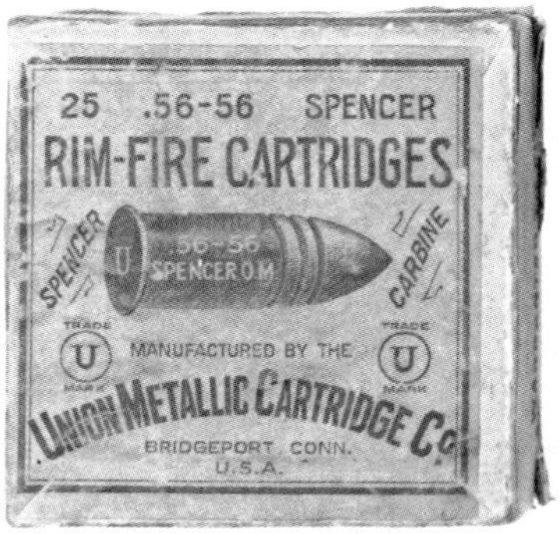
25 .56-56 SPENCER
RIM-FIRE CARTRIDGES
SPENCER
CARBINE
56-56
SPENCER O.M.
MANUFACTURED BY THE
UNION METALLIC CARTRIDGE Co
BRIDGEPORT, CONN.
U.S.A.

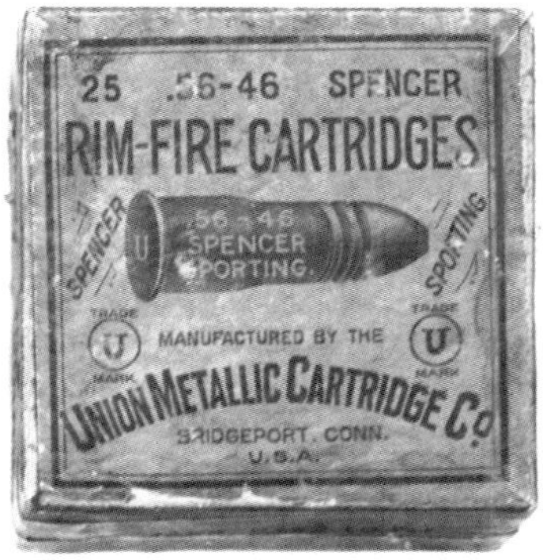
SPENCER
RIM-FIRE CARTRIDGES
SPORTING
SPENCER
SPORTING
MANUFACTURED BY THE
UNION METALLIC CARTRIDGE Co
U.S.A.

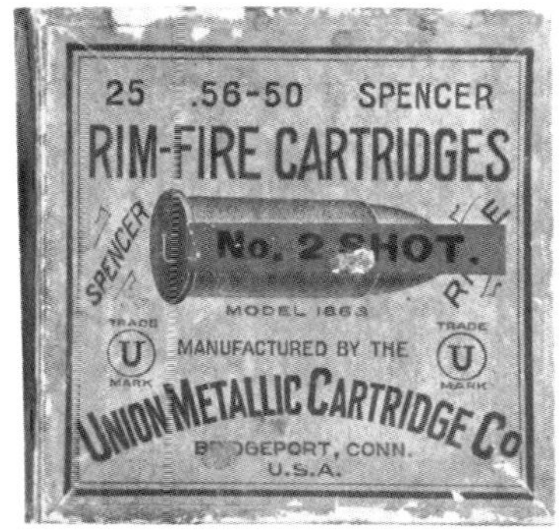
25 .56-50 SPENCER
RIM-FIRE CARTRIDGES
SPENCER
RIFLE
MODEL 1863
MANUFACTURED BY THE
UNION METALLIC CARTRIDGE Co
CONN.
U.S.A.

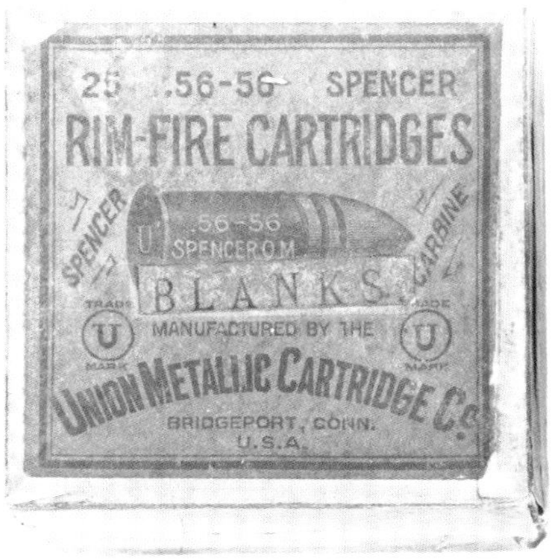
SPENCER
RIM-FIRE CARTRIDGES
SPENCER
CARBINE
56-56
SPENCER O.M.
MANUFACTURED BY THE
UNION METALLIC CARTRIDGE Co
U.S.A.

25
U.M.C.
PRIMED
Metallic Cartridges
FOR
SPORTING
RIFLE
U.S.A.

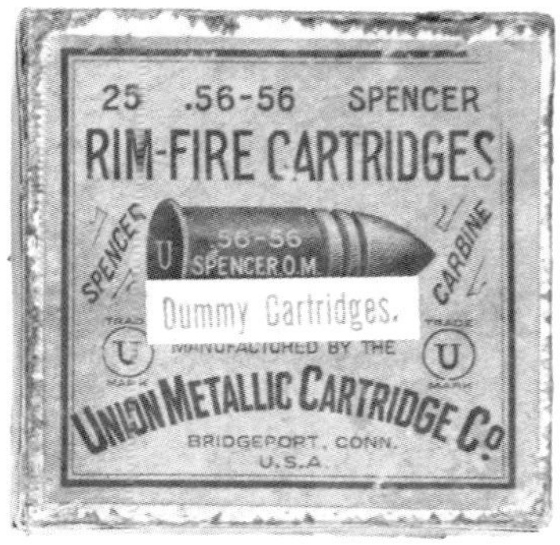
25 .56-56 SPENCER
RIM-FIRE CARTRIDGES
CARBINE
56-56
SPENCER O.M.
Dummy Cartridges.
MANUFACTURED BY THE
UNION METALLIC CARTRIDGE Co
BRIDGEPORT, CONN.
U.S.A.

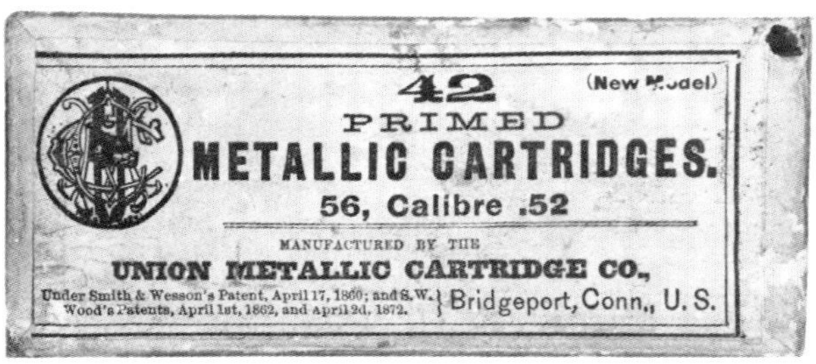
42
PRIMED
METALLIC CARTRIDGES.
56, Calibre .52
MANUFACTURED BY THE
UNION METALLIC CARTRIDGE CO.,
Bridgeport, Conn., U. S.

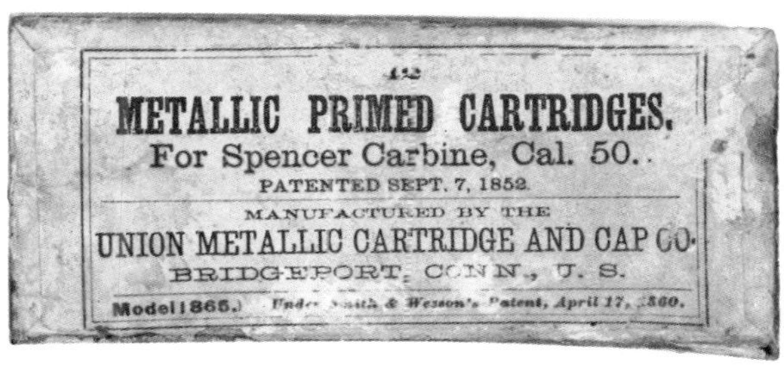
METALLIC PRIMED CARTRIDGES.
For Spencer Carbine, Cal. 50.
PATENTED SEPT. 7, 1852.
MANUFACTURED BY THE
UNION METALLIC CARTRIDGE AND CAP CO.

42
(Model 1863)
PRIMED
METALLIC CARTRIDGES.
56, Calibre .50
MANUFACTURED BY THE
UNION METALLIC CARTRIDGE CO.,
Bridgeport, Conn., U. S.

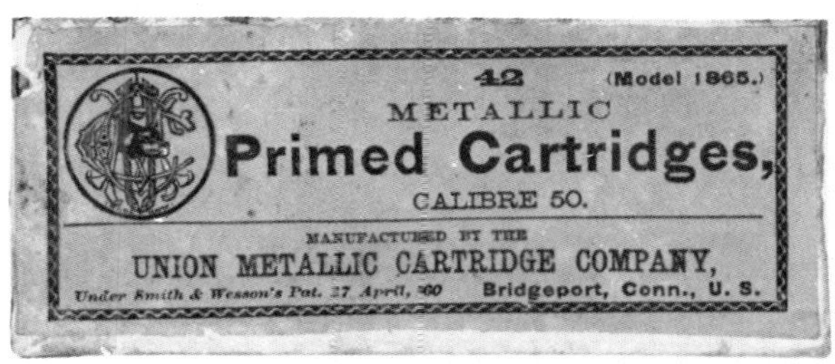
42
(Model 1865.)
METALLIC
Primed Cartridges,
CALIBRE 50.
MANUFACTURED BY THE
UNION METALLIC CARTRIDGE COMPANY,
Bridgeport, Conn., U. S.

APPENDIX B

Cartridge collecting dates back as early as the 1850's. Lt. C.M. Wilcox, United States Army, and Maj. Rudolf Schmidt, Swiss Army, were among the earliest to study ammunition and write about it. The Frankford Arsenal collection assembled for display at the Philadelphia International Exposition of 1876, covered in a publication edited by the late Col. B.R. Lewis and issued by the Smithsonian Institution, is a valuable source of information on U.S. as well as European ammunition up to that date. The extensive collection and notes of General John R. Pitman housed in the library of the West Point Military Academy is also a very important source, but yet unpublished.

Cartridge collecting as a hobby gained impetus with the publication of Herschel C. Logan's illustrated "Cartridges" in 1948. The National Cartridge Collectors Association was formed in the United States in the early 1960's and was soon retitled the International Cartridge Collectors Association.

Interest soon spread throughout the world and resulted in the formation of the European Cartridge Collectors Club as well as the other associations listed in the following page. Information is disseminated through regular publications to the members.

The unofficial annual meeting of the ICCA occurs in the Chicago, Illinois USA area during April, and the ECCC sponsors an annual show in September, usually in Switzerland or Luxembourg. These social and trade gatherings are attended by collectors from several continents. The ECCC has Dutch and German national groups which also sponsor annual meetings. New Zealand's club, the South American groups, and the state or regional associations in the U.S. also conduct regular trade and display award shows.

Joining one or more of the collector associations is the most profitable way of advancing one's interest in the hobby. Individuals dealing in collector cartridges are listed in the trade section of GUN DIGEST annual. Books of interest, as well as sources, are also listed there.

International Cartridge Collectors Association, Membership Chairman Victor v.B. Engel, 1211 Walnut St., Williamsport, PA 17701 USA. Issues a 30-page magazine every other month, giving coverage to all aspects of the hobby. Dues schedule and other information available from the above-named individual. Interested persons in other countries may arrange an associate membership through their own collector organization, if they so desire.

European Cartridge Collectors Association, c/o Mr. Emile Timmermans, Stw, Op Brussel 493, B-1900, Overijse, BELGIUM. Issues a monthly question-and-answer bulletin. New members should indicate preference for English, French, or German edition. Information also available from the above individual concerning the Dutch and German collector associations.

New Zealand Cartridge Collectors Club, c/o Mr. Graham Lewis, Secretary-Treasurer, 10 Reid Road, Wainuiomata, NEW ZEALAND. Issues a monthly news letter.

Uruguayan Association of Collectors of Weapons and Cartridges, c/o Dr. Richard G. Rimini, Calle Canelones 1285, Montevideo, URUGUAY

Argentinian Association of Collectors of Weapons And Cartridges, c/o Mr. T. Dalle Nogare, P.O. Box 28, SUC 1 q B., Buenos Aires, ARGENTINA

Ohio Cartridge Collectors Club, c/o Mr. Russ Pidala, Secretary-Treasurer, 16108 Ashland Drive, Brookpark, Ohio 44142 USA

Texas Cartridge Collectors, c/o Mr. Frank Chalfant, Secretary, 16003 El Camino Real, Houston, Texas 77062 USA

Oregon Cartridge Collectors, c/o Mr. Richard King, 3228 N.W. 60th, Corvallis, Oregon 97330 USA

Kansas Cartridge Collectors Assn., c/o Mr. Douglas Seneke 6720 Fontana Prairie Village, Kansas 66208 USA

Sioux Empire Cartridge Collectors Association, c/o Mr. Bern Schliesman, 3108 South 4th Ave., Sioux Falls, South Dakota 57105 USA

California Cartridge Collectors Association, c/o Mr. James E. Estes, 1943 W. Houston, Fullerton, California 92622 USA

Alberta Cartridge Exchange, c/o Mr. Nick Krevosheia, 6619-137th Ave., Edmonton, Alberta, CANADA T5C-2L1

Cartridge Collectors Association of Edmonton, c/o Mr. Guenter Wienert, 16415-79A Ave., Edmonton, Alberta, CANADA T5R-3J2

BIBLIOGRAPHY

BARTLETT, W.A. & D.B. GALLATIN

1878. Cartridge Manual. Lists every cartridge patent issued in the United States, England, and France prior to 1878. Reissued by Pioneer Press, Harriman, Tenn., no date.

BLANCH, H.J.

1909. A Century of Guns. London.

CAPITAINE, EMILE & VON HERTLING, Ph FREIHERR

1888. Die Kreigswaffen.

DATIG, FRED A.

1967. Cartridges for Collectors, Vol. III. Los Angeles.

DOLLECZEK, ANTON

1896. Monographie der k.u.k. österr.-ung. Blanken und Handfeuerwaffen. Austria. Reprinted in 1970 by Akademische Druk- u. Verlagsanstalt, Graz, Austria. Detailed descriptions of military arms of Austria and Austria/Hungary from the matchlock era to 1896, two plates of ammunition drawings.

EDWARDS, WILLIAM B.

1962. Civil War Guns. Reissued by Castle Books, Secaucus, N.J., 1978. Full illustrated historical review of the small arms used by the North and South during the American Civil War, 1861-1865.

FREMANTLE, HON. T.F. (later Lord Cottesloe)

1901. The Book of the Rifle. London.

FIGUIER, LOUIS

N.D. Les Merveilles De La Science.

FULLER, CLAUDE E.

1933. The Breech-loader in The Service. Topeka, Kansas.

———

N.D. The Rifled Musket. New York. Bonanza reprint.

GUNTHER, DR. REINHOLD
1909. Allgemeine Geschichte der Handfeuerwafen. Leipzig.

GREENER, W.W.
1899. The Gun and Its Development.

HULL, EDWARD A.
1978. Providence Tool Co. Military Arms. Milton, Florida. Covers the contracts for muzzle-loading arms received by the company in 1861 through the Peabody and Peabody Martini. Illustrated.

LEWIS, COL. BERKELEY R.
1956. Small Arms and Ammunition in The United States Service. Washington. The definitive work on U.S. small arms and ammunition, gunpowder manufacture, and accouterments from pre-revolutionary days through the Civil War. Numerous plates devoted to ammunition.

____
1959. Notes on Ammunition of the American Civil War. Washington.

____
1972. Small Arms Ammunition at The International Exposition, Philadelphia, 1876. Washington. The Frankford Arsenal collection assembled and described by Lt. Henry Metcalf for the exposition. Fully illustrated.

LOGAN, HERSCHEL C.
1948. Cartridges. Huntington, W. Virginia. Several reprintings. A pictorial digest of small arms ammunition from paper through centerfire. A standard work.

LONGSTAFF, MAJ. F.V. AND ATTERIDGE, CAPT. A. HILLIARD
1917. The Book of the Machine Gun. A historical review as well as detailed study of the arms used by the armies of WW I. Numerous plates.

LUGS, JAROSLAV
1956. Firearms Past and Present. Prague, Czechoslovakia. Reprinted in English in 1973, London. A broad historical review of firearms development with a separate accompanying text of photo illustrations and drawings. Some valuable information on ammunition development.

MAJENDIE, V.D. & C.O. BROWNE
1870. Military Breech-loading Rifles. Woolwich. Reissued by Arms and Armour Press, London, in association with Fortress Publications, Inc. P.O. Box 241, Stoney Creek, Ontario, Canada, 1973.

MATTENHEIMER, A.
1868. Die Patronen der Rücklandungsgewehre (Cartridges for Breech-loading Arms). A Morion reprint by Intersico Press, AG, Zurich, Switzerland, 1975. Contains two color lithograph foldout plates containing 163 actual size drawings of cartridges, bullets, and rifle actions of Europe and America to that date, together with descriptions of each in text.

METSCHL, JOHN
1928. The Rudolph J. Nunnemacher Collection of Projectile Arms, Vol. I, Long Arms. Bulletin Vol. 9, Public Museum of Milwaukee, Wisconsin.

MORDECAI, MAJOR ALFRED
1860. Military Commission to Europe in 1855 and 1856. Washington. Includes Capt. Julian Schön's "Rifled Infantry Arms," 1855, translated from the German by Capt. J. Gorgas, United States Army, with 13 plates detailing ammunition, projectiles, arms, etc. of European countries.

MYATT, MAJOR F.
1979. The Illustrated Encyclopedia of 19th Century Firearms. New York.

NORTON, BREVET BRIGADIER GENERAL CHARLES B.
1872. American Breech-loading Small Arms. New York.

OMMUNDSEN, H. & E.H. ROBINSON
1915. Rifles and Ammunition. New York and London.

POLLARD, MAJOR HUGH BERTIE CAMPBELL
1926. A History of Firearms. Boston and New York.

RILING, RAY
1951. Guns and Shooting. New York. Bibliography on the subject commencing with "Das Feuerwerksbuch von 1420" (Fireworks Manual of 1420), through 2,769 listings to 1949.

SCHMIDT, MAJOR RUDOLF (Later Colonel)
1875. Die Handfeuerwaffen. Basel. Contains 56 color plates of small arms, exploded drawings, ammunition, and detailed descriptions of each. Reprinted 1968 by Akademische Druk- u. Verlagsanstalt, Graz, Austria, combining the 1875 and a revised 1878 edition with 76 plates in the original color.

________, RODOLPHE
1877. Les Armes A Feu Portatives. Paris. French edition of the above work.

SCHNEIDER, HUGO; AM RHYN, MICHAEL; KREBS, OSCAR; REINHART, CHRISTIAN; SCHIESS, ROBERT
1970. Bewaffnung und Ausrüstung der Schweizer Armee seit 1817. Handfeuerwaffen System Vetterli (Armament and Equipment of the Swiss Army from 1817. Small Arms System Vetterli), Zurich. Contains color plate and technical details of 20 variations of 10.4mm Swiss rimfire ammunition. Exhaustive coverage of the weapons.

SCHNEIDER, HUGO & AM RHYN, MICHAEL
1979. Continuation volume of above. Eidgenössische Handfeuerwaffen. Zurich. Color plate and technical details on primitive paper and rimfire ammunition and illustrated coverage of the arms from flintlock through the rimfire transformed muzzle-loading rifles.

SCHÖN, CAFT. JULIAN

1855. Rifled Infantry Arms. Dresden. Translated from the German by Capt. J. Gorgas, United States Army from the 2nd edition, augmented. Included as an appendix to Major Mordecai's "Military Commission to Europe—", which see.

SELLERS, FRANK

1978. Sharps Firearms. North Hollywood, California

SHARPE, PHILIP B.

1938. The Rifle in America.

STOCKBRIDGE, V.D.

1874. Digest of U.S. Patents Relating to Breech Loading and Magazine Small Arms. Reissued 1963 by Norm Flayderman. Abridged description of each of the original patent illustrations, together with an index of patentees' names. Over 800 patents covered.

R.Q. SUTHERLAND and R.L. WILSON

1971. The Book of Colt Firearms. Kansas City, Missouri. Encyclopedic coverage of the company's arms from the beginning, illustrated in color and black and white.

SUYDAM, CHARLES R.

1960. The American Cartridge. Santa Ana, California. The definitive work on the rimfire catridge. Second edition revised, 1973.

————————

1977. U.S. Cartridges and Their Handguns. North Hollywood, California. Cartridge development from the handgun point of view.

SWENSON, G.W.P.

1972. Pictorial History of the Rifle. New York.

THIERBACH, MORITZ

1886 and 1899. Die Geschichtliche Entwicklung der Handfeuerwaffen (The Historical Development of Small Arms). Dresden. Reprinted with original color plates 1965 by Akademische Druk- u. Verlagsanstalt, Graz, Austria. Recognized as one of the finest works on the subject. The plates illustrate ammunition types from the earliest times in actual size, and the text provides descriptions in detail.

VON ELLGER, KARL

1868. Kriegsfeuerwaffen. Leipzig. A Morion reprint in two volumes, 1978, by Intersico Press, AG, Zurich, Switzerland.

VON PLÖNNIES, W.

1865. Das Zundnadel-Gewehr. Darmstadt & Leipzig. A Morion reprint by Intersico Press AG, Zurich, Switzerland, 1976. Fully illustrated study of the Prussian needle rifle and its ammunition.

WAHL, PAUL & TOPPEL, DON

1965. The Gatling Gun. New York.

WALTER, JOHN

1979. The German Rifle.

WEYGAND, HERMANN

1872. Infanterie-Präzisionswaffen. A Morion reprint by Intersico Press, AG, Zurich, Switzerland, 1978. Detailed discussions and illustrations of arms and cartridges from the needle rifle to the early centerfires adopted by European countries.

WILLIAMSON, HAROLD F.

1952. Winchester—The Gun That Won The West. Washington.

WILSON, LT. C.M

1859. Rifles and Rifle Practice. New York. Detailed analyis of early rifle experiments, drawn for the most part from French military reports. Several plates, including one of cartridge drawings showing ammunition in use by European countries and United States to that time. See Lewis' work of 1956 for the same plate.

WINANT, LEWIS
1959. Early Percussion Firearms. New York. Reprint.

GOVERNMENT AND MISCELLANEOUS ANONYMOUS PUBLICATIONS

Danske Orgelespingoler Med Enhedspatroner in the Tøjhusmuseet Skrifter 3, Kobenhaven, 1945.

Die Entwicklung der Gewehrmunition in der Schweiz, von 1815-1945—Eidg. Munitionsfabrik, Thun, Switzerland, n.d.

Report of the Board on Behalf of United States Executive Departments at the International Exhibition, Philadelphia, Pa., 1876, Vol I, 1884. Contains descriptions, illustrations, and lists of the ordnance section, including listing of the small arms ammunition display prepared by Lt. Henry Metcalf.

Treatise on Ammunition, 1897. Artillery College, Woolwich, England.

Catalogue Firearms Collections, 191-. United States Cartridge Company.

PERIODICALS

The American Rifleman, magazine of the National Rifle Association, U.S.A., various issues.

The Cartridge Collector, publication of the European Cartridge Collector Club, editor Emile Timmermans. Various issues.

The Gun Report mazazine, various issues June 1955 to the present. Cartridge editor Frank Wheeler, succeeded by Louis Behling.

The International Cartridge Collector, publication of the International Cartridge Collectors Association, editor John A. Belton.

INDEX